AGENTS OF CHAOS

Earthquakes, Volcanoes, and Other Natural Disasters

Stephen L. Harris

Mountain Press Publishing Company
Missoula, Montana
1990

Copyright © 1990
Stephen L. Harris

Cover Painting by
Dorothy Sigler Norton

Library of Congress Cataloging-in-Publication Data

Harris, Stephen L., 1937-
    Agents of chaos : earthquakes, volcanoes, and other natural
disasters / Stephen L. Harris.
      p. cm.
    Includes bibliographical references and index.
    ISBN 0-87842-243-9 : $12.95
    1. Earthquakes—West (U.S.) 2. Volcanoes—West (U.S.)
3. Natural disasters—West (U.S.) I. Title.
QE535.2.U6H37 1990
551.2' 0978—dc20   90-6165
                                     CIP

Mountain Press Publishing Company
P.O. Box 2399 • Missoula, MT 59806
(406) 728-1900

Dedication

For Karen, Will, Kevin, Dan, and Steve.

Preface

"To see a glacier, witness an eruption and feel an earthquake . . ." so G. K. Gilbert described the "natural and legitimate ambition" of a "properly constituted geologist." Dr. Gilbert, a former head of the United States Geological Survey, may have applied these goals equally to anyone interested in the forces that shape our earth.

By great good luck I grew up in western Washington state where mountain glaciers are abundant, earthquakes almost guaranteed every decade, and volcanoes like Mt. St. Helens periodically stage awe-inspiring eruptions. Privileged to live in a region that is still geologically young and restless, I have been able to experience intimately some of nature's most spectacular forces. During the last few decades I have climbed over, under, and inside active glaciers, scaled icy mountaintops to camp inside steaming craters, shuddered with the thrill of the earth trembling beneath my feet, and observed at close range a death-dealing volcano.

Many other residents of the Far West have also personally experienced one or more of these phenomena. Yet most Americans are unaware that such events as great earthquakes or explosive volcanic eruptions are not isolated happenings, but part of a vast ongoing process that involves the little understood force of chaos. The sudden injection of violent change into a previously stable environment, chaos operates intermittently in all our lives. The 1980s began with the cataclysmic eruption of Mt. St. Helens, witnessed the deaths of 25,000 people in Columbia from the Nevado del Ruiz volcano, and ended in 1989 with Alaska's Mt. Redoubt belching ashclouds that temporarily halted all air traffic in our largest state. Redoubt's outburst followed only two months after northern California was shaken by the most costly earthquake in North American history. The United States government recognizes that tens of millions of Americans, not only on the Pacific Coast, but in the Midwest and on the Atlantic seaboard as well, face geologic violence they are now ill-prepared to handle. Noting that giant earthquakes, volcanic eruptions, or devastating sea waves and floods are inevitable, the U.S. Geological Survey and the Smithsonian Institution have declared the 1990s to be the International Decade of Natural Disaster Reduction.

Agents of Chaos explains where and why Americans can expect their lives to be disrupted by natural disasters and what they can do to mitigate their effects. The book answers questions about where, when, and how often volcanoes will blaze and great earthquakes will suddenly rearrange the landscape. It pinpoints the dangers and demonstrates what is likely to happen when the next catastrophic quake strikes California, Alaska, the Pacific Northwest, the central Mississippi Valley, or the East Coast. It also explains why many of Mt. St. Helens' sister volcanoes, now deceptively quiet, will erupt, threatening towns and popular resorts in eleven different states.

From ancient American Indian legends to the latest scientific research, this book incorporates a wide variety of sources to present a new perspective on the power of chaos and the geologic hazards through which it operates. My debt to the many earth scientists whose work contributed to this volume is gratefully acknowledged in the bibliographies at the end. Among those who offered valuable suggestions are Brian Atwater of the U.S. Geological Survey, University of Washington, Seattle; Joseph Ziony, Assistant Director at the California Division of Mines and Geology, Sacramento; and Dave Alt and Don Hyndman of the University of Montana, Missoula. I am also grateful to David R. Montgomery, who permitted use of his photos of the 1989 California earthquake; Mary Woods, editor of California Geology, at the California Division of Mines and Geology, Sacramento; Mike Moore, the helpful librarian of the U.S. Geological Survey's photo library in Menlo Park, California; David Hirsh, photo curator at the U.S. Geological Survey's Tacoma office; and Rick Hayes, a talented artist at the University of California, Davis, whose maps and drawings vividly illustrate the geological processes discussed.

Table of Contents

Part I:
Earthquake Hazards in the United States

Chapter 1

AGENTS OF CHAOS

A clear day, predicted to be sunny and calm, suddenly turns cloudy and thunderous. The heart of a previously healthy runner abruptly flutters and stops. Children happily playing in a schoolyard are gunned down by a psychopathic intruder. Without warning, the seemingly solid earth beneath our feet begins to heave and roll like a sea in storm, toppling buildings and killing thousands of people.

These unexpected disruptions of the normal order are manifestations of chaos, the sudden injection of violent change into a previously stable environment or system. As scientists are increasingly aware, chaos plays a major and, as yet, an entirely unpredictable role in both geologic processes and human lives. From wild fluctuations of the stock market to earthquakes, volcanic eruptions, and the behavior of subatomic particles, chaos permeates the very fabric of the universe.

Although chaos plays its tricks everywhere, its presence in geologic events is particularly evident along the Pacific rim, where most of the world's great earthquakes and nearly eighty percent of its volcanic activity occur. All parts of the United States experience earth tremors, but the states bordering the Pacific Ocean, including Alaska and Hawaii, enjoy far more than their fair share of natural disasters.

The general public is aware that damaging earthquakes frequently strike California, typically causing tens or hundreds of millions of dollars in property losses. Only recently, however, have earth scientists begun to realize how enormous is the potential for chaotic events in other western states. Only during the late 1980s have physicists, astronomers, and earth scientists recognized that chaos, extraterrestrial as well as geological in origin, has created widespread havoc that suddenly—and violently—changed the face of North America.

A large meteorite striking near the southeastern corner of Oregon about 17 million years ago probably triggered the floods of molten basalt lava that formed the Columbia River Plateau and also generated the concentrated heat source, a hot spot, that now underlies Yellowstone National Park. Some earth scientists also believe that another meteorite, plunging into the Pacific millions of years earlier, ruptured the oceanic crust to create the Hawaiian hot spot from which lava has steadily erupted to build the Hawaiian Islands.

We can scarcely imagine the appalling effects of the tsunamis, huge seismic sea waves, that raked the Pacific coastline when the meteorite struck near the present site of Hawaii. The waves must have been thousands of times the size of the tsunamis that drowned 36,000 persons during the 1883 eruptions of Krakatau, a volcanic island in Indonesia. In this century, merely terrestrial events, the 1964 Alaska and 1960 Chile earthquakes, generated massive tsunamis that pounded the shores of Hawaii and the United States mainland. The Alaskan quake sent waves up to twenty feet high slamming against the coasts of Alaska, Washington, Oregon, and northern California, killing 122 people.

Current studies indicate that the West Coast may experience even more catastrophic earthquakes and tsunamis. Sand and mud deposits along the coasts of Washington, Oregon, and northern California show that large areas of the coastline have undergone rapid sinking and submergence, repeatedly, during the last few thousand years. A growing number of earth scientists believe that the Pacific Northwest may expect gigantic earthquakes, greater than any that have shaken California during historic time. Equal to the strongest upheavals recorded in this century, such superquakes would cause widespread coastal subsidence and flooding that could devastate many low-lying towns and cities.

As the U.S. population steadily increases, even minor earthquakes claim a heavy economic toll. The tremor that hit the California town of Whittier, near Los Angeles, on October 1, 1987, registered a mere 5.9 on the Richter scale, but it killed seven persons and caused $215 million in property damage. Two years later, the 7.1 magnitude Loma Prieta earthquake in northern California was even more financially devastating, with damage estimates ranging from $7 to $10 billion, the costliest geologic event ever to occur in North America. Destructive as they were, the Whittier and Loma Prieta temblors are relatively small potatoes compared to the great earthquake that is expected to convulse the Los Angeles or San Francisco regions sometime during the next few decades. Geologists foresee a repeat of the Richter magnitude 8.3 earthquake that rolled through southern California in

1857, an event that released a thousand times more energy than the recent Whittier shake.

Although large earthquakes strike the West more often, some of the most powerful quakes in United States history shook the Mississippi Valley. A series of high magnitude jolts centered near New Madrid, Missouri, in 1811-1812 was felt over a million square miles, from Boston to New Orleans. Some earth scientists predict that eastern states, including New York, South Carolina, or those in the central Mississippi Valley will experience a catastrophic shock by the year 2000. Because East Coast quakes cause severe shaking over a much larger area than those in the West and because the East is less well prepared to deal with major earthquakes, damage and loss of life will be much greater proportionately than in California's dreaded "Big One."

In 1980 it took Mount St. Helens only a few minutes to devastate more than 200 square miles of southwestern Washington, killing nearly sixty persons and transforming evergreen forests into a gray, lifeless wasteland. Geological processes commonly operate with painful slowness, but St. Helens demonstrated that a large natural environment can be changed with deadly swiftness.

St. Helens has many sister volcanoes in the Cascade Range, which stretches northward from California to British Columbia, some of which are potentially as explosive as she. Eleven have erupted during historic time and some may revive soon. Of the many prospective sites of future eruptions in the western United States, the Mono Lake-Long Valley area in east-central California is perhaps the most threatening. The numerous earthquakes centered in the area, as well as ground uplift and an increase in hot spring and other thermal activity, suggest that a significant eruption may occur there in the near future.

As if high magnitude earthquakes, destructive tsunamis, and recurring explosive eruptions in Alaska, the Cascades, and California's Mono Lake area were not enough to worry about, Americans must also consider the kind of volcanic cataclysms that formed the Yellowstone caldera, the vast collapse depression in which Yellowstone National Park sits. Three times during the last 1.8 million years, the Yellowstone volcano has raged with an explosive violence that far exceeds any eruption known to human history. The last great paroxysm ejected towering waves of incandescent ash that swept over much of the western and central United States. Each of the three climactic eruptions was spaced about 600,000 years apart. Because the last occurred about 600,000 years ago and because a large reservoir of molten rock still simmers beneath Yellowstone, many geologists expect the volcano to produce another catastrophic eruption.

The Yellowstone caldera is only one of many that pockmark the western terrain, from California's Long Valley to Valles caldera in New Mexico to Creede caldera in Colorado. These and many others produced eruptions of extraordinary power and volume, devastating thousands of square miles. Although such colossal events occur infrequently, a repetition of these caldera-forming eruptions would be not a regional but a national catastrophe that would change large areas of North America almost beyond recognition.

The greatest chaotic events, such as meteorite impacts and the large-scale volcanic eruptions they cause, are mercifully rare, occurring perhaps every 26 to 30 million years. They devastate the entire planet and probably help account for the fact that ninety-nine percent of all life forms that ever existed are now extinct. Powerful earthquakes or volcanic eruptions occur far more frequently, striking some part of the U. S. almost every decade. Millions of westerners will probably witness a destructive volcanic eruption, while many millions more, from New York to Seattle, will experience a cataclysmic earthquake. Although random, chaotic events are inevitable, striking with a sudden violence that will change forever the lives of countless Americans.

Chapter 2

EXPERIENCING AN EARTHQUAKE

The quake started with a distant low rumble, then hit with the impact of a Mack truck slamming full speed into the side of a house. One moment our sixth grade class squirmed, awaiting release into the noon-hour sunshine of a mild April day. The next, we crowded toward the door in near panic as the floor lurched beneath our feet and plaster showered down upon our heads. Shock waves traveled visibly up brick walls as windowpanes rattled and cracked with a deafening roar. Outside the second-story classroom, looking up at the stairwell ceiling, I saw the school's massive outside wall separate from adjoining partitions and jerkily sway back and forth, spewing chunks of mortar on teachers and children scrambling down the staircase. Shivering and creaking throughout its three stories, Tacoma's venerable Edison Elementary School audibly protested the earthquake's assault on its Victorian dignity.

Our class reached the outdoors safely that April 13, 1949. Other students were not so fortunate. Across town an eleven-year-old patrol boy was crushed to death by bricks tumbling from his school's fourth-story dormer as he emerged for lunch-hour duty. A young companion was badly injured. As in California's Long Beach earthquake of 1933, school buildings were particularly hard hit. Throughout the Puget Sound area, thousands of chimneys toppled, roofs fell in, and masonry walls crashed into the streets, killing several other people and seriously injuring scores more. The quake was felt over 150,000 square miles of the Pacific Northwest, left eight dead, and caused over $25 million in property damage.

The 1949 quake's epicenter, the point on the earth's surface directly above the earthquake's underground source, was near Olympia, Washington state's capital, where nearly every large structure was damaged. The capitol's Roman-style dome was conspicuously cracked, while its graceful stone cupola had to be replaced—by the ugly metal facsimile that caps the building today. From the Canadian boundary to south of the Oregon state line, fallen chimneys, cracked ground, rockslides, and crumbled masonry marked the destructive passage of what is still the most severe earthquake to jolt the Pacific Northwest during historic time.

Registering a magnitude of 7.1 on the Richter scale, the temblor was a "major" quake, not a "great" earthquake like the 8.3 shock that shattered San Francisco in 1906. Although less well known, the 1949 event was as powerful as the 1989 earthquake that devastated parts of the San Francisco Bay Area, and considerably stronger than the quakes that wreaked havoc in the San Fernando Valley in 1971 and caused $215 million worth of damage near Los Angeles in 1987.

The 1949 earthquake was also strong enough to stimulate my lifelong fascination with the geologic forces that can so dramatically convulse the earth, injecting an unpredictable element of excitement and danger into our lives. Walking home from school that clear April noon, I remember gazing at the immense glacier-clad bulk of Mount Rainier that dominates Tacoma's eastern horizon and wondering if there could be any connection between earthquakes such as I had just experienced and the chain of volcanic peaks—including Rainier and St. Helens—that stretches from British Columbia to northern California. Even then some earth scientists had noted the suggestive proximity of earthquake-prone regions and the belt of volcanoes that encircles the Pacific Ocean—the notorious "Ring of Fire."

Today geologists can explain why people living along the Pacific rim, whether in Japan, Indonesia, Chile, California, or the Pacific Northwest, seem to experience more than their fair share of the world's earthquakes and volcanic eruptions. Earthquakes occur in virtually all fifty of the United States, but most of the high magnitude shocks center in the West. Although California's earthquakes attract the most publicity, other less populous states, such as Alaska, are shaken more frequently—and commonly by much stronger tremors. Recent studies indicate that the Pacific coast areas of Washington and Oregon may be at even higher risk than California for superquakes of almost unimaginable destructiveness.

The more earth scientists learn about the physical evidence left by large earthquakes and volcanic eruptions during the recent geologic past, the clearer it becomes that such events are an integral part of the western scene. These geologic expressions of chaos have played a major role in creating the present landscape and will continue to reshape it in the future. As the populations of California, the Pacific

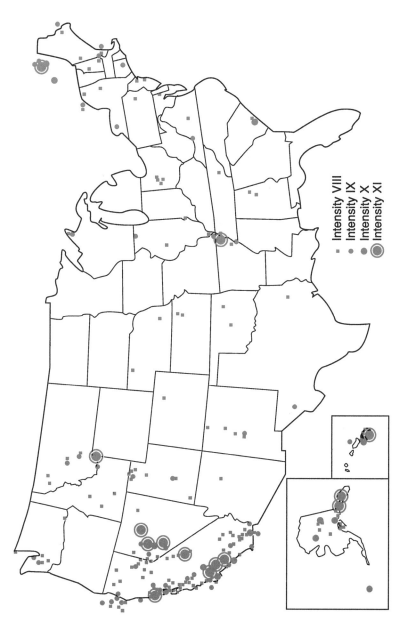

Intensity VIII
Intensity IX
Intensity X
Intensity XI

This map illustrates the geographical distribution of large, damaging earthquakes that have shaken the United States during historic time, the last 300 years. Note that extremely strong quakes, with a Modified Mercalli intensity of X or above, have repeatedly struck the Far West, particularly Alaska, California and Nevada. Although fewer great earthquakes strike east of the Rockies, their damaging effects extend over a much wider area and thus represent a greater hazard to larger numbers of people.

Northwest, and Alaska steadily increase, earthquakes and volcanoes will also have an ever-growing public and economic impact. St. Helens' 1980 eruptions were small compared to possible future activity at some other western volcanoes, but they caused estimated short-term losses to the economy of Washington state of $970 million. A decade later, the accumulated costs were still mounting!

Minor or moderate earthquakes occur much more often than great ones, but they serve to illustrate what havoc California's anticipated "Big One" may inflict. When moderate earthquakes center near urban centers, they cause economic losses disproportionate to their size. Besides taking seventy lives, the 1971 San Fernando and 1987 Whittier earthquakes together caused more than one billion dollars in damage. With economic losses nearly ten times higher, the 1989 earthquake in northern California was the most expensive natural disaster in United States history. Even that record loss will be shattered by the great earthquake that seismologists expect to hit near Los Angeles and/or San Francisco during the next few decades.

Geologists regard such damaging earthquakes and volcanic eruptions as inevitable. Public awareness of these geologic hazards, including why, where, and how often they occur, is essential. As Americans regularly cope with destructive storms, hurricanes, tornadoes, floods, and crippling blizzards, so we must also learn to deal with the potential threat of killer earthquakes and volcanoes.

What Causes an Earthquake?

An earthquake is the sudden trembling or shaking of the ground caused by the abrupt movement or displacement of rock masses within the earth's crust. Most earthquakes originate within the upper ten to twenty miles of the lithosphere, the earth's rigid outer shell. Powerful forces in the lithosphere exert stress on the rock, pushing or pulling it. Rock is elastic enough to accumulate strain, bending or changing shape and volume. When stress exceeds the strength of the rock, the rock breaks along a preexisting or new fracture plane called a fault. The fracture rapidly extends outward from its place of origin, the focus. As the rock breaks, waves of energy—seismic waves— radiate through the earth, causing the vibrating and shaking of an earthquake.

Seismic waves, generated by friction and crushing as masses of rock slide past one another, travel outward from the earthquake focus like ripples on a pond. The fastest are the primary, or P, waves, which compress the rock in front of them and elongate it behind as they rush through the planet at about three to four miles per second. Next come

Primary or P wave

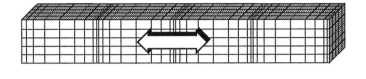

Secondary or S wave

Love wave

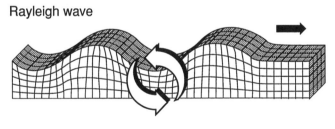

Rayleigh wave

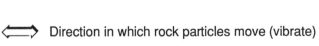

⟸⟹ Direction in which rock particles move (vibrate)

➡ Direction of travel of waves

Earthquakes generate three basic kinds of seismic waves. Primary, P, waves travel fastest, compressing rock in front of them and elongating it behind. Secondary, S, waves cause an undulating up-and-down movement. The most damaging to structures are surface waves, Love and Rayleigh waves, that shake the ground both vertically and in a zig-zag motion.

the S, or secondary, waves, which undulate, causing an up-and-down and side-to-side motion as they roll through at about two miles per second.

Most damaging to man-made structures are the surface waves, called Rayleigh and Love waves, which shake the ground both vertically and horizontally. They can create a visible rolling and billowing of the surface as well as a jerky zig-zag motion that is particularly destructive to high-rise buildings. The slowest moving of seismic waves, surface waves cause the worst devastation because they accelerate ground motion and take longer to travel through a given area.

The degree of damage to buildings and landscapes depends largely on their proximity to the epicenter and the nature of the underlying soil. In general, damage is most severe within a few tens of miles of the earthquake source and diminishes with increasing distance. Certain kinds of soil, however, increase the intensity of destructive shaking, even many tens of miles from the epicenter. In 1989 San Francisco's Marina District, although nearly sixty miles from the point at which the San Andreas fault ruptured, suffered extensive damage. The Marina soils, sandy landfill covering an old lagoon, underwent partial liquefaction. A secondary effect of earthquakes, liquefaction occurs when water-saturated sands or silts shake violently; water pressure forces sand grains apart, causing the subsurface materials to act as a liquid. Transformed into mush, liquified sediment can flow laterally underground, cracking the ground surface and allowing heavy buildings to tilt or sink into the liquified muck. The syrupy soil may break in waves, as it did in the Marina, or erupt in geysers of liquid sand. The liquefaction of unconsolidated soils with a high water table is responsible for some of the worst damage in many earthquakes, including those that hit Alaska in 1964, Mexico City in 1985, and the Bay Area in 1989.

America's most celebrated source of large earthquakes, the San Andreas fault slices through California's coastal mountains from the Mexican border to Cape Mendocino, where it passes into the ocean floor. The San Andreas marks the boundary between two immense slabs of crustal rock that are slowly grinding past each other. The slab or tectonic plate holding the Pacific Ocean basin creeps northwestward at the rate of an inch or two per year, rubbing against the western edge of the North American plate. When the two crustal plates become temporarily locked together, strain accumulates until the plates break apart, releasing seismic energy that sets the Earth vibrating. Large-scale movement along two different sections of the 700-mile-long San Andreas fault has triggered two of the greatest earthquakes in United States history, those of 1857 and 1906. The east and west sides of the San Andreas changed their relative positions only a few feet during the 1989 quake, but in 1906 they shifted as much as twenty-two feet.

The underground rupture initiating an earthquake will continue until it reaches areas in which the rock is not sufficiently strained to permit it to extend farther. The 1906 earthquake ruptured the northern San Andreas for a distance of 280 miles, most of the length marked by visible surface fracturing. Despite considerable secondary cracking of the ground surface in the Santa Cruz Mountains and elsewhere, the 1989 quake did not produce any detectable primary surface rupture.

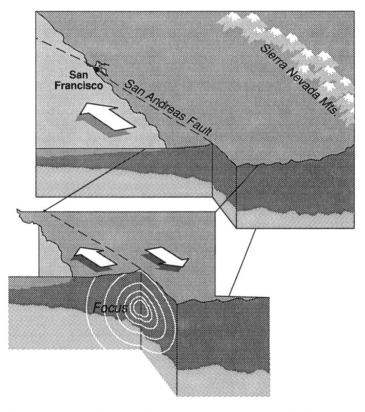

The San Andreas fault marks the boundary between the North American and Pacific Ocean plates. When the two plates are temporarily locked together, strain accumulates until the snagged rock masses fracture and the Pacific Plate lurches northwestward. In the Loma Prieta earthquake of 1989 a thirty-mile long segment of the fault shifted about five feet. The focus lay 11.5 miles beneath the Santa Cruz Mountains, roughly fifty-six miles south of San Francisco.

Besides causing damage through fault movement, violent shaking, soil liquefaction, ground failure, and landslides, earthquakes may also trigger tsunamis, seismic sea waves that can travel thousands of miles across oceans at speeds of 300 to 400 miles an hour. As the tsunamis approach shore, they begin to drag on the seafloor, slow down, and rise to heights of fifty feet or more. Such earthquake-produced waves have repeatedly slammed into low-lying coastal areas in Hawaii, Alaska, and the Pacific Northwest.

The numerous earthquakes and dozens of volcanic eruptions that occur annually around the Pacific rim are not unusual. These bursts of violence that terrorize people in Tokyo, Tacoma, San Francisco, or Mexico City are merely demonstrations that earth's processes operate chaotically, pulsating with sudden change and movement.

The Modified Mercalli Scale

In contrast to the magnitude scales, which measure the strength of an earthquake, seismologists use an intensity scale to describe the severity of earthquake effects in a given area. Unlike the Richter magnitude, which is determined objectively by recording earthquake wave motion on a seismograph, classification of intensity depends on subjective observation of the earthquake's effects on persons, buildings, and the ground surface at a particular place. The most commonly used rating scale is the Modified Mercalli Scale of 1931, which distinguishes twelve degrees of earthquake severity. Examples (paraphrased and abridged):

VI. Felt by everyone, many run outdoors, general excitement. Heavy furniture may be shifted; some cracked or fallen plaster, damaged chimneys....

VII. Everyone runs outside. Damage negligible in buildings of good design and construction; slight to moderate in well-built ordinary structures; considerable in poorly built or badly designed structures. Some chimneys felled. (Equivalent to VIII on the Rossi-Forel Scale.)

VIII. Most people frightened; many approach panic. Water, sand, mud ejected from the earth in small to moderate quantities. Ground surface and pavement cracked. Damage considerable in ordinary substantial buildings and severe in poorly built structures. Fall of chimneys, factory stacks, columns, monuments, walls. Changes in water levels of streams and wells.... (Equivalent to VIII+ to IX on the Rossi-Forel Scale.)

IX. Great slumping, fissuring, uplift and other deformities of the ground surface. Few, if any, masonry structures remain standing. Bridges and dams destroyed. Underground pipelines completely disrupted and unfunctional. Tsunami (seismic sea) waves sweep up coastlines; large quanitites of water, sand, and mud erupt from the earth as geysers....

XII. Damage total, with nearly all structures severely damaged or destroyed. Ground disturbances great, with large fissures, shearing cracks. Landslides and rockfalls extensive; slumping and collapse of river banks. Avalanches dam lakes, divert rivers, produce waterfalls. Waves seen on ground surfaces. Lines of sight and level distorted. Objects thrown into the air....

Measuring an Earthquake

Earth scientists have devised several means of determining the size, power, intensity, and location of earthquakes. A seismograph is the instrument used to detect and record seismic wave amplitude. An international network of seismograph stations is maintained all over the planet, so that within minutes of a large earthquake seismographs at many widely scattered locations record the seismic waves it produces. Because the different kinds of seismic waves travel at different speeds, they are picked up by a seismograph in an established order. Although all three types originate from the earthquake focus at the same time, the P waves arrive first, the S waves next, and the Surface waves third. The time intervals between the first arrivals of the P, S, and Surface waves increase with distance from the focus. By analyzing the time lapse between consecutive arrivals of the three seismic wave types, earth scientists can calculate the approximate distance from the seismograph station to the earthquake source. By comparing the records at several different stations, seismologists can also determine the quake's location.

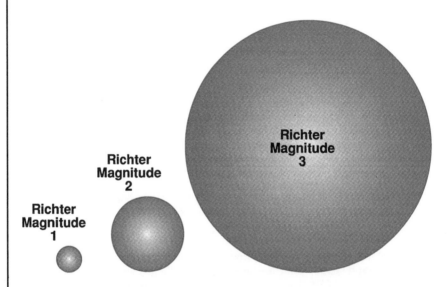

Relationship between earthquake magnitude and energy. Notice the exponential growth in size and power for each increase of a whole number in the Richter scale for measuring earthquake magnitudes. The volumes of the spheres are roughly proportional to the amount of energy released by earthquakes of the magnitudes given. Using the same scale to represent the energy released by the San Francisco earthquake of 1906 (Richter magnitude 8.3) would require a sphere with a diameter of 220 feet!

The Richter scale, which seismologist Charles F. Richter devised in 1935, is commonly used to calculate the magnitude or amount of energy an earthquake releases. A numerical scale that assigns numbers in ascending order of magnitude, from 0 to 8.6, the Richter scale is logarithmic. The difference between two consecutive whole numbers on the scale means an increase of ten times in the size of ground vibration. Earth scientists estimate that this tenfold increase in seismic wave amplitude requires an increase of about 31.5 times in energy release. Thus the 7.1 magnitude earthquake that struck northern California in 1989 was fully thirty-one times more powerful than the 5.9 shock centered in Whittier, near Los Angeles, in 1987. The great San Francisco earthquake of 1906, Richter magnitude 8.3, was more than a thousand times stronger.

Three "great" earthquakes (magnitudes 8.0 or above) have shaken California during the last 130-odd years: the Fort Tejon, near Los Angeles, in 1857; the Owens Valley, east of Yosemite National Park, in 1872, and the San Francisco in 1906. At least two Alaska quakes, in 1899 and 1964, were significantly larger.

Because the Richter scale becomes saturated at about the 8.3 magnitude, seismologists have recently adopted a new means of measuring the energy release of exceptionally strong earthquakes.

According to the Moment Magnitude scale, introduced in 1977 by a geophysicist named Kanamori, the 1964 Alaska quake (8.4 on the Richter scale) actually had a reading of 9.2—the largest ever recorded in North America. The 1960 earthquake in Chile was even greater—M 9.5—the highest yet measured anywhere in the world.

An earthquake rated between about 7.0 and 7.9 on the Richter scale is classed as a "major" seismic event. Major shakes, such as the 7.1 Puget Sound quake of 1949, are more frequent than great ones and have taken many lives and caused millions of dollars in damage. An event comparable to the 7.1 magnitude quake that killed sixty-seven northern Californians in 1989 strikes the Golden State on the average of once every eighteen years.

Some of the West's most damaging earthquakes rank as only "moderate" in seismological terms. Registering between about 6.0 and 6.9 on the Richter scale, some moderate shocks have been centered near densely populated areas where they caused far greater loss of life and property than much stronger ones that occurred in largely uninhabited areas. Two of California's moderate quakes, in Long Beach (1933) and in the San Fernando Valley (1971), together caused more deaths than any U. S. earthquake since 1906.

Chapter 3

PLATES IN MOTION:
Our Dynamic Earth

Viewed from outer space, Earth is a shimmering blue-green sphere streaked with white clouds. From the moon, an observer can clearly discern the familiar outline of the pale continents, separated by expanses of dark ocean. If we were to go back in time 200 million years, however, the earth's face would look very different. The Atlantic Ocean did not exist then, and North and South America were parts of a vast landmass called Pangaea surrounded by an almost unbroken sea.

Shortly after 200 million years ago, Pangaea began to break up, driven by the same subterranean forces that move the continents today. A great rift, a linear zone of deep fractures in the earth's crust, split the giant continent. As the two sides of the rift pulled away from each other, forced apart by the upwelling of molten rock from below, a large basin began to form and fill with water—the primitive Atlantic Ocean. The slow opening of the Atlantic, a few inches per year, gradually separated South America from Africa and North America from Europe.

Like giant barges, the continents thus began their majestic progress across the earth's surface. Their present positions on the global map are only temporary. In a few more million years, cartographers will have to redraw the world map as North America drifts westward and fractions of the continents split from the main landmasses to pursue their own individual journeys. The Red Sea is a modern rift zone that will continue to widen, further separating Africa from Asia. The westernmost coast of California is in the act of divorcing itself from the rest of the Golden State and heading northwestward for a

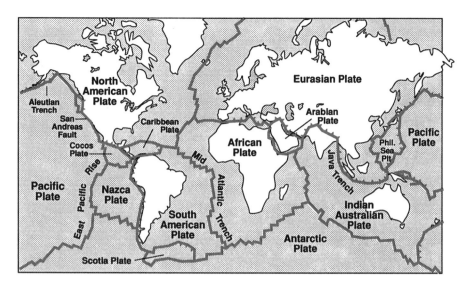

Like the cracked shell on a hardboiled egg, the earth's crust is broken into several large and numerous small segments or plates. Powered by dynamic forces within the earth, the crustal plates are in constant motion, spreading apart (diverging) or clashing together (converging). Most of the world's earthquakes and volcanic eruptions occur in linear zones along plate boundaries, here marked in red.

rendezvous with Alaska. The present Hawaiian Islands will eventually cease to be tropical as they glide northward toward their ultimate submersion in the Aleutian trench.

The continents, which to most people seem comfortably solid and permanent, are neither stable nor stationary. Throughout earth's five billion year history, they have probably been constantly on the move—fleeing from each other, colliding together, and again pulling apart to follow their individual itineraries. The ocean basins have had shorter lives than the continents; they have opened wide only to close again or be gobbled up as ocean floors sank beneath encroaching continents and were consumed in the earth's hot interior.

The relatively new theory of plate tectonics can explain much about why the continents move, creating earthquakes and volcanoes in the process. The earth's rigid outer crust—thinner proportionately than the skin on an apple—is a brittle and easily fractured cover on the planet's incandescent interior. Earth scientists believe that the planet is composed of three major zones. A tremendously hot but solid inner core of heavy metals like iron and nickel is surrounded by a molten outer core. Between the core, which is about the size of the moon, and the outer crust lies a broad zone of hot plastic rock called the mantle, approximately 1800 miles thick. The upper mantle and solid rocky crust form the third zone, the lithosphere.

The lithosphere is broken into about a dozen major slabs or plates, making the earth's surface a kind of jigsaw puzzle. Some of the plates, covered with oceanic crust composed of dense volcanic rock called basalt, hold ocean basins. Other plates carry the continents on their backs. Because they are composed of a lighter granitic material, the continents "float" higher on the underlying rock of the mantle than do the denser oceanic plates.

Underwater volcanic eruptions of basalt lava are constantly creating new ocean floor. Most of these basaltic eruptions occur along linear fracture zones and build long ranges of submarine volcanic mountains. The Mid-Atlantic Ridge, which marks the rift at which the Atlantic Ocean basin first opened, is such a range. Fresh effusions of molten rock, or magma, rise from depth to erupt within the rift, creating new ocean floor as opposite sides of the rift zone move apart. The same eruptive process proceeds all along the ocean ridge that wanders across the Pacific Ocean floor, the trend of its axis resembling the curved seam on a baseball.

Oceanic ridges constantly create new crustal material, thrusting ocean floors away from the linear eruption zones, called spreading centers. Along the rim of the Pacific basin, old ocean floor pushed outward from the submarine spreading centers is being destroyed as it sinks under the margins of the bordering continents. In the collision of oceanic and continental plates, the heavier basaltic ocean floors sink, or are subducted, beneath the ocean-facing edges of the continents. The subduction process, in which the ocean plate descends into the upper mantle, typically in fits and starts, generates the thousands of earthquakes that occur every year in Alaska, Japan, Indonesia, western South America, Mexico, and the northwestern United States.

Subduction is also responsible for the linear belts of volcanoes that encircle the Pacific rim. As the water-saturated ocean floor sinks under its own weight into the relatively soft mantle, the descending rocky slab is heated. Water boils out of the sinking slab as hot steam and rises into the overlying edge of the continental plate. The addition of water to the subterranean hot rocks lowers their melting temperature and generates new magma. Pools of magma form reservoirs of molten rock underground. Gas-rich and hotter than the surrounding mantle rock, the magma rises until it erupts on the surface. Thus chains of active volcanoes form all along the Pacific rim where subduction is carrying the old oceanic lithosphere under the margins of the continents.

These circum-Pacific volcanoes include some of the world's most famous mountains, such as Fujiyama in Japan, Shishaldin and Pavlof in Alaska, and Rainier and Shasta in the Cascade Range. In contrast to the oceanic volcanoes like those in Hawaii, which erupt fluid lava flows in a relatively quiet manner, the continental volcanoes tend to be violently explosive. Volcanoes in Alaska and the Cascades are thus particularly dangerous to human life.

Two relatively small sections of the Pacific Ocean floor, the Juan de Fuca and Gorda plates, are now sinking under the western edge of North America at the rate of two to three inches per year. The Juan de Fuca and Gorda plates descend along the Cascadia subduction zone, which extends about 800 miles from Cape Mendocino, California to central Vancouver Island, British Columbia. Their movement generates earthquakes and also provides the molten rock that fuels St. Helens and its fellow Cascade volcanoes. Far to the north, the much larger northern Pacific plate descends more rapidly in a subduction zone paralleling the Alaskan Peninsula and the Aleutian Islands. Because the oceanic lithosphere sinks more swiftly and on a larger scale beneath Alaska than along the west coast of the United States, Alaska experiences scores of potentially damaging earthquakes every year. Next to Indonesia, it also produces more volcanic eruptions annually than any other place on earth.

As the following chapters demonstrate, no spot in the West is absolutely free from earthquake or volcanic hazards. Some places, however, including certain parts of Alaska, Hawaii, California, and the Pacific Northwest are outstandingly vulnerable to geologic disasters. Inescapably caught up in the earth's relentless movements, some areas will almost certainly experience a major natural calamity before the end of this century.

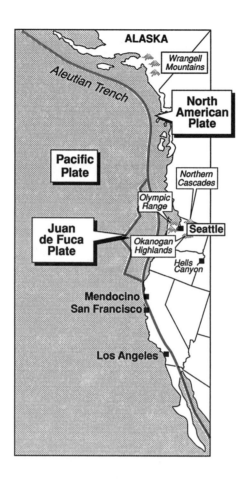

Eastward movement of the Pacific plate served as a conveyor belt that transported ancient oceanic islands and microcontinents to the western edge of North America. Between about 150 and 40 million years ago, a series of foreign or "exotic" terranes rafted to and fused against the continental coast, including previously equatorial microcontinents now forming the Okanogan Highlands and North Cascades of Washington, as well as numerous segments of Idaho, Oregon, and California. Westward-drifting North America eventually overrode most of the original east Pacific subduction zone, leaving only a remnant, the Juan de Fuca plate, operating today.

Chapter 4

THE FAR WEST IS A GEOLOGIC CRAZY QUILT

"A land of shreds and patches!" This misquotation of the Bard aptly suggests the crazy-quilt makeup of western North America. The Far West is not an original part of the continent, but consists of bits and pieces of earth's crust that originated far from their present locations and were later grafted onto the western margin of North America.

A traveler zigzagging the length of California today may drive over as many as fifty different land fragments known as exotic or foreign terranes. In contrast to the familiar word "terrain," denoting a geographical region or the topographical features of a certain tract of land, "terrane" refers to a geologic entity that differs markedly in both composition and history from neighboring areas.

A terrane does not have the same family history as the surrounding landforms. It is literally an outlander, a stranger of foreign birth that has somehow migrated to a geographical region with which it has little in common.

The West is a mosaic of exotic, foreign terranes, a jigsaw puzzle of distinct crustal slabs and segments that formed thousands of miles away from their present location. Most of California, the Pacific Northwest, British Columbia, and Alaska are collages of broken crustal fragments that originated somewhere in the Pacific Ocean and then traveled vast distances to join our continent.

Their fossils of ancient plant and animal life not native to North America show that these terranes are indeed alien. In the long process of reaching their present position, the terranes were commonly split up, their fragments rotated and sent traveling in different directions. Visitors to Idaho's Hells Canyon see formations that came from some unknown point in the equatorial Pacific and may belong to the same geologic sequence that now forms the lofty Wrangell Mountains in Alaska, 1200 miles distant.

Many of the discrete terranes that have been crushed and glued to our western shores were once islands probably similar in size and appearance to modern Japan, Indonesia, New Guinea, New Zealand, and others in the western or equatorial Pacific. Some were large enough to be called "microcontinents," like the one that docked fifty million years ago to become the rugged North Cascades in north-central Washington.

Washington, in fact, is composed almost entirely of crumpled microcontinents and other exotic terranes. Before about 100 million years ago, the Pacific Ocean covered all but the extreme northeastern corner of the state. The first traveling landmass to come ashore was the Okanagan microcontinent, which now forms the highland region between the Columbia and Okanagan rivers in northeastern Washington, and extends northward into British Columbia.

Following the arrival of the North Cascades microcontinent, the western part of the state was formed chiefly of old oceanic crust. The North Cascades landmass probably imported its own subduction zone, which still functions off the western Washington coast. Oceanic crust attached to the western margin of that microcontinent now forms the bedrock of the Willapa Hills of southwestern Washington. The rugged Olympic Mountains are composed of rock that was formerly stuffed into the off-shore subduction trench and has since risen through the slab of basaltic oceanic crust that forms the Willapa Hills.

The notion that apparently solid and stable landmasses have moved vast distances around the globe, colliding and merging to create our western landscape, may seem incredible, but it is a natural consequence of the earth's plate movements. Like giant rafts, the plates carry continents, islands, and ocean basins on their backs as they travel over the hot, plastic material of the earth's mantle.

The ocean floors, growing outward from the underwater spreading centers in the oceanic ridges have acted as conveyor belts, transporting oceanic islands and isolated microcontinents to an inevitable encounter with the leading edge of North America. Piece by piece, once autonomous landforms smashed against the granite wall of the continent, creating a coastline that is a complex accretion of exotic terranes.

The Far West grew as the Pacific plate dived under North America, scraping off thick deposits of seafloor sediments against the continental coastline as it descended, adding new building blocks to the western terrane.

Most of the coast ranges of California, Oregon, Washington, and part of British Columbia formed when lighter sediments and slabs of basalt, originally emplaced underwater, broke free of the descending plate. These lighter rock masses, compressed and folded, rose along the continental margin to form today's coastal hills and mountains.

Most of the recognized exotic terranes were imported to the West during late Mesozoic and early Cenozoic time, between about forty and 150 million years ago. Although growth by accretion of ocean islands has now ceased in the Far West, Earth's rocky plates continue to shift position, rearranging the global map. Large and small landmasses are even now traveling toward distant destinations, ultimately to create some surprising geographical combinations.

Much of the California coast, for example, is headed for Alaska. Having long ago overridden an ancient subduction zone, California south of Cape Mendocino no longer has the Pacific sea floor sinking beneath it. Instead, the Pacific plate, including that part of California lying west of the San Andreas Fault, is grinding north-northwestward alongside the western edge of North America. After relocating the site of Los Angeles in the suburbs of San Francisco, which sits on the continental side of the fault, continued plate movement will eventually transport southern California's once sunny beaches to Alaska.

Another western state is also destined for relocation. The Hawaiian Islands, sitting atop the Pacific plate, are creeping northwestward at a rate of about four inches per year. Eventually the island flotilla will be swallowed in the Aleutian trench, a deep submarine trough that parallels Alaska's Aleutian Islands and Russia's Kamchatka Peninsula, and marks the northern boundary of the Pacific plate. Islands now flourishing in tropical splendor beneath blue skies in time will sink into the Aleutian subduction zone, crushed in the incandescent interior of the earth.

If they escape remelting in the subduction mill, some island fragments may become embedded in formations along the Aleutian Island chain. As late twentieth century geologists discovered exotic terranes in our Far West, so geologists of the distant future may find remnants of a tropical paradise amid the landscapes of chilly Alaska.

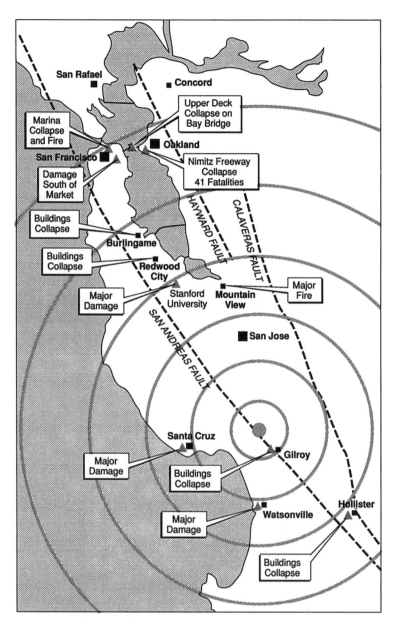

The 7.1 magnitude Loma Prieta earthquake of 1989 was the largest earthquake in northern California since 1906, and the largest anywhere in the state since 1952. Damage was most extensive within a radius of about twenty miles from the epicenter, but seismic waves traveling through unconsolidated land fills in San Francisco and Oakland, about sixty miles to the north, caused Mercalli shaking intensities of IX, devastating the Marina district and parts of west Oakland, including the Nimitz Freeway, part of Interstate 880. Sixty-seven people were killed and official damage estimates ranged up to $10 billion, making this quake the most costly natural disaster ever to occur in North America.

Chapter 5

THE CALIFORNIA
EARTHQUAKE OF 1989

For fifteen terrifying seconds San Francisco's Candlestick Park swayed and shuddered. Chunks of concrete fell from cracking walls and stairways, but the 60,000 baseball fans packing the stadium for the third game of the World Series remained generally calm, even after officials cancelled the event. No one yet had any idea how deadly the brief shaking was.

Only gradually did people become aware that a major disaster had struck northern California at 5:04 p.m., October 17, 1989. Fans leaving the ballpark and rush hour commuters found numerous streets and highways suddenly closed, creating traffic jams many miles long. A column of smoke and flame expanded rapidly over San Francisco's Marina district, reflecting the glow of an autumn sunset, while other fires smoldered in the East Bay. As night deepened, almost the entire Bay region lay in darkness, punctuated by endless lines of car headlights and the red flare of police, fire, and other emergency vehicles. In many areas it would be days before electrical power and communications were restored.

Operating on emergency power reserves, a few local television and radio stations slowly gathered the news: a fifty foot section of the San Francisco-Oakland Bay Bridge had collapsed, killing at least one person and severing the city's main transportation link with the outside world. The Embarcadero Freeway along San Francisco's waterfront was severely damaged and landslides and quake-weakened bridges had closed Highways 1 and 101 to the south. The worst devastation occurred on Interstate 880 in Oakland, where a 1.5-mile-long section of the Nimitz Freeway's upper deck had collapsed onto the lower deck, crushing or trapping an unknown number of travelers. More than forty bodies were later recovered from the wreckage.

The next morning brought reports that towns south of San Francisco had suffered even more extensive damage. The seaside resort of Santa Cruz, home to the University of California's most picturesque campus, was effectively isolated, with highways 17, 9, and 1 blocked by rockfalls or damage to bridges. In Santa Cruz County scores of old brick buildings crumbled and hundreds of frame houses were pitched off their foundations, leaving several thousand people temporarily homeless. Many expensive houses in the Santa Cruz Mountains were reduced to kindling wood, while others, more or less intact, slid several yards downhill.

Measuring 7.1 on the Richter scale, the Loma Prieta temblor was the strongest to hit northern California since the 1906 San Francisco earthquake. Although centered in the Santa Cruz Mountains roughly fifty-six miles south of San Francisco, the quake created havoc

Partly collapsed house in the town of Los Gatos, California. From Santa Cruz to San Francisco, damage to wood-frame buildings was extensive. In Watsonville, near the Loma Prieta epicenter in Santa Cruz County, approximately one out of every eight homes was destroyed, throwing hundreds of people into shelters and tent villages. —California Division of Mines and Geology, David R. Montgomery photo

A pressure ridge caused surface buckling and cracking at this parking lot near the Santa Cruz Boardwalk. Sand and water also erupted nearby and at spots along the shores of San Francisco Bay, sixty miles distant.

throughout the Bay Area. Seismic waves traveling outward from the epicenter on the San Andreas fault created intense shaking in landfill and mud along the margins of San Francisco Bay. Water-saturated soils temporarily liquefied, causing extensive fracturing, slumping, and sliding of the ground surface. The Marina district, built on sand fill deposited after the Panama Pacific Exhibition in 1915, was the most severely affected. Thirty-five buildings totally collapsed and about 150 others had serious structural damage. Some poorly constructed frame apartment houses telescoped, bringing their third or fourth floors level with the street. Ignited by a ruptured gas main, fires ravaged several blocks of the Marina, reenacting on a smaller scale the quake-triggered conflagration that destroyed most of the city after the 1906 earthquake. In both events San Francisco firefighters were hindered by lack of water resulting from broken mains. In 1989 neighborhood volunteers dragged fire hoses many blocks to pump water from the Bay, but it took most of the night to bring the blaze under control.

Most of the thousands of other buildings damaged in San Francisco also stood on landfill. Areas south of Market Street and the Mission district suffered pockets of heavy damage, particularly at the sites of buried sloughs or streambeds where the ground rippled like jello. Five people were killed at Sixth and Bluxome streets when a brick facade toppled into the street. Thousands of others were injured by glass from shattering windows and various falling objects.

The full horror of the disaster was concentrated in west Oakland at a segment of Interstate 880 known as the Cypress structure. Observers said that the quake rolled along the two-decker freeway like a colossal ocean wave behind which section after section of the upper deck collapsed. Scores of drivers along a 1.5-mile stretch of the lower level were crushed beneath tons of concrete and steel that compressed their vehicles to a thickness of eighteen inches. Clouds of dust rising from the shattered freeway and fallen walls of nearby buildings mingled with the stench of burning tires, spilled gas, and oily smoke.

Near collapse, this apartment house is one of scores of buildings wrecked in San Francisco's Marina district. Numerous structures collapsed altogether, while others were so badly cracked or thrown out of allignment that they had to be torn down immediately. —California Division of Mines and Geology, David R. Montgomery photo

28

This view of the Cypress Viaduct clearly shows that the inadequately reinforced concrete columns supporting the upper deck were sheared off, causing it to collapse. Scores of cars, trucks, and their occupants were crushed by the pancaking of the two decks. —California Department of Transportation, Bob Colin photo

More than forty people died when a 1.5-mile-long section of the upper deck of Interstate 880, known as the Cypress Viaduct, collapsed onto the lower deck. —California Department of Transportation, Bob Colin photo

A truck driver and his rig were in the right place at the wrong time heading north along Interstate 880. With countless tons of concrete and steel crashing down around it, this truck survived relatively unscathed. Most travelers on this stretch of freeway in west Oakland were not so fortunate.
—California Department of Transportation, Bob Colin photo

Collapsed section of the San Francisco-Oakland Bay Bridge. —California Department of Transportation, Bob Colin photo

Unconsolidated land fill at the Port of Oakland liquefied during the 1989 earthquake, causing ground slumping and surface cracking at this parking lot. —California Division of Mines and Geology, David R. Montgomery photo

Despite fears that aftershocks would bring down the remaining structure, many ordinary citizens living near the collapsed freeway immediately risked their lives to rescue drivers trapped in what had become a gigantic concrete tomb. Later fire crews and other professionals labored for days to retrieve the dead. Few on the lower deck survived; one man found alive after eighty-nine hours subsequently died of his injuries.

The national media focused on spectacular damage to Bay Area bridges, freeways, and the Marina, but less visible damage to other structures was also significant. Numerous masonry churches, fine old houses, hotels, municipal buildings and other historic edifices were so seriously weakened that they will have to be torn down or undergo enormously expensive repairs. One of the quake's saddest legacies may be the loss of some of northern California's most distinguished architecture.

31

The Cooper House, a Santa Cruz landmark originally built in 1867, suffered extensive structural damage in the Loma Prieta earthquake and had to be demolished. —Annie Kappl photo

This partly collapsed brick building in the downtown Santa Cruz mall is typical of thousands of structures destroyed throughout Santa Cruz county, where thousands of people were left homeless. —California Division of Mines and Geology, David R. Montgomery photo.

Boards support earthquake-damaged brick walls in downtown Santa Cruz. —Annie Kappl photo

Lateral ground spreading and sagging along the San Andreas fault zone, Hazel Dell Road, Santa Cruz County, California.
—U.S. Geological Survey,
M. Rymer photo

This is particularly true in areas closer to the epicenter, where elegant old buildings in Santa Cruz, Los Gatos, Watsonville, Hollister, Gilroy, and San Jose either collapsed or experienced structural damage. Three people were killed by falling walls in Santa Cruz's Pacific Garden Mall, a six-block cluster of about 100 shops housed in Victorian era brickwork. Scores of downtown buildings, wrecked beyond repair, were quickly demolished, causing a resident to note that with the vanishing of irreplaceable architecture, "Santa Cruz has lost its soul."

The Loma Prieta quake, named for a peak in the Santa Cruz Mountains near its epicenter, produced a highly irregular distribution of damage. While the University of California at Santa Cruz, a mere ten miles southwest of the epicenter, experienced only minor losses, Stanford University in Palo Alto suffered an estimated $165 million in damage. The sprawling megalopolis of San Jose had only scattered damage, while San Francisco and Oakland, thirty to forty miles farther from the quake center, had most of the fatalities. Even in the Bay region severe damage was generally confined to isolated pockets like the Marina district and west Oakland, although other

This shattered chimney and fireplace is typical of earthquake damage to unreinforced masonry. Note that the wooden house experienced little structural damage.
—California Division of Mines and Geology, David R. Montgomery photo

34

Although cracking of the ground surface was common in the epicentral region of the Santa Cruz mountains, geologists were unable to find any signs that the San Andreas fault had ruptured the surface. View of Summit Road area near Highway 17, Santa Cruz County, California. —U.S. Geological Survey, M. Rymer photo

Part of Highway 101 collapsed over Struve Slough near Watsonville, where damage was extremely severe. Many older bridges and overpasses suffered partial or total collapse, but highways designed to withstand strong ground shaking generally performed well. The newly constructed Highway 92/101 interchange in San Mateo has styrofoam inserts that separate adjoining sections of the concrete deck, allowing them to readjust during violent shaking. The earthquake separated some expansion joints by up to several inches, but the cloverleaf held together. —California Division of Mines and Geology, David R. Montgomery photo

San Francisco neighborhoods, such as the Mission, Richmond, Sunset, Haight, and Tenderloin, also had significant losses. Much of the Bay Area went largely unscathed.

As in previous California earthquakes, well-built structures erected on solid rock suffered little or no damage, while badly constructed wood frame and unreinforced masonry buildings fared poorly. The worst damage occurred either near the epicenter or on landfill where liquefaction presumably took place.

The Loma Prieta quake was triggered by movement along the San Andreas fault at a depth of about 11.5 miles, where the crustal block west of the faultline shifted northward about five or six feet. No visible surface faulting occurred, although violent shaking opened numerous superficial fractures over an area sixty miles long and twenty-five miles wide. The largest such crack, located in the Santa Cruz Mountains near the intersection of Summit Road and Highway 17, was 650 yards long and 2.5 yards wide. Judging by the zone along which hundreds of aftershocks occurred, the subterranean Loma Prieta rupture is about thirty miles long. This compares to the 280-mile-long surface rupture that accompanied the 1906 earthquake.

Although the most expensive natural disaster in United States history, causing damages estimated at between seven and ten billion dollars, the Loma Prieta earthquake was merely a bitter foretaste of the Big One expected to devastate California within the next thirty years. Earth scientists anticipate a magnitude 8.0 or larger event, at least thirty times more powerful than the 1989 tremor. Whether it strikes at the northern or southern segments of the San Andreas fault, it will probably resemble one of the great historic quakes described next.

Chapter 6

CALIFORNIA'S DEADLY EARTHQUAKES

The vertical boundary dividing North America from the Pacific Ocean basin, the San Andreas fault is a giant fracture cutting through approximately 700 miles of California's coastal ranges. During the last thirty million years, the crustal block on the Pacific side of the fault has spasmodically lurched and slid several hundred miles to the north. Movement along the fault is irregular because the adjoining continental and oceanic plates commonly lock together until enough tension accumulates to break them apart, generating the seismic waves that produce most of California's largest and most costly earthquakes. In 1906 the San Andreas shifted up to twenty-two feet, triggering violent shaking that killed at least 3000 people in and around San Francisco. The Loma Prieta quake of 1989 was considerably less powerful, but it caused the highest dollar losses of any natural event in American history.

Society's collective memory is short, and most Americans, even those living on or near major fault zones like the San Andreas, do not know where, how often, or with what force earthquakes have struck the areas they inhabit. By surveying some of California's damaging earthquakes, we can learn which localities in our most populous state are most at risk. Seismologic and geologic evidence can also give us a fair idea of where to expect major quakes in the future. Probability forecasts are particularly important, because quakes equal to or larger than the Loma Prieta event shake California on the average of once every eighteen years.

Three great earthquakes, all of which had estimated Richter magnitudes of 8.0 or higher, struck California during one fifty-year period, in 1857, 1872, and 1906. Scores of less powerful quakes, centered near

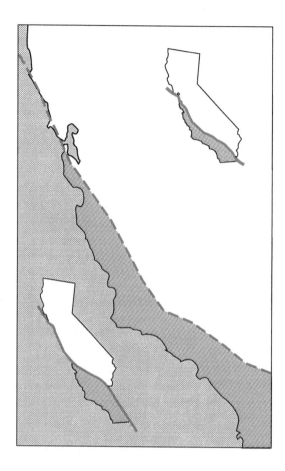

The most notorious earthquake zone on earth, the San Andreas Fault slices through California from the Mexican boundary northward to Cape Mendocino, where it passes into the ocean floor. Marking the contact between two huge slabs of the earth's crust, the Pacific and North American plates, the San Andreas generated two great earthquakes during one fifty year period, the Fort Tejon in 1857 and the San Francisco in 1906. Earth scientists anticipate another massive shock on the southern San Andreas, probably in the Los Angeles region, within the next few decades.

urban areas, have also caused widespread damage. Although stronger earthquakes have occurred in other states, notably Alaska, California's temblors have exacted higher dollar losses and caused more than twice the number of deaths than the combined totals of earthquake effects in the other forty-nine states. As California's population has grown, so have the lethal consequences of its earthquakes.

Fort Tejon, 1857

On January 9, 1857, part of the southern San Andreas Fault broke near Tejon Pass, offsetting the earth's surface between the San Bernardino Valley and Cholame Valley, a distance of more than 225 miles. It was the mightiest shock to afflict southern California since Europeans first settled there. No seismographs existed in 1857, so we will never know the exact magnitude of the earthquake. Modern estimates place magnitude at about 8.3. The Fort Tejon quake caused severe shaking from south of Los Angeles northward to San Francisco and Sacramento, a wide area then sparsely inhabited. Although every building was leveled at Fort Tejon, only one fatality resulted—a woman who was killed when her adobe house collapsed. Today, more than twelve million persons live within forty miles of this part of the fault, and tens of thousands of casualties and billions of dollars of property damage could result from a similar earthquake.

San Francisco, 1906

At 5:12 A.M., Wednesday, April 18, 1906, the northern San Andreas fault ruptured from San Juan Bautista to Point Arena, creating visible surface displacement over most of that 280-mile distance. The most powerful upheaval to occur within the forty-eight coterminous states during this century, the San Francisco earthquake caused tremendous destruction far beyond the city itself. Numerous coastal towns and villages suffered heavy damage, including Santa Rosa and San Jose. Many of the massive Romanesque buildings on the Stanford University campus disintegrated into heaps of rubble, killing two students. One hundred persons died when the poorly built Agnews Hospital for the insane almost totally collapsed. Locally along the coastline, high cliffs plummeted into the Pacific, the earth split in

The domed library of Stanford University, about thirty-five miles south of San Francisco, rises above the rubble of other campus structures demolished by the 1906 earthquake. Although most of Stanford's imposing Romanesque buildings collapsed, only two students were kiled by the early-morning upheaval. —U.S. Geological Survey, J.C. Branner photo

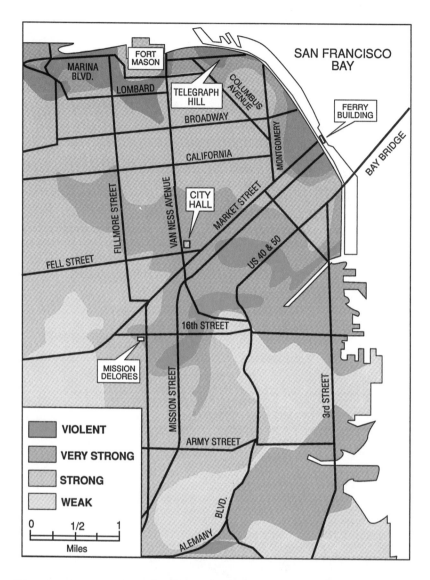

*Map showing variation of shaking intensities in San Francisco during the
1906 earthquake. Much of the worst damage occurred in areas underlain
by landfill, particularly along the waterfront and in the South of Market
and Mission districts, where hundreds of wooden and masonry buildings
totally collapsed. Red and dark pink shades identify areas of extensive
liquefaction, surface cracking, and ground failure. The Marina and South
of Market neighborhoods also suffered heavy damage in the 1989 quake.*

gaping fissures, railroad cars were thrown from their tracks, and rails were contorted into grotesque shapes. South of Santa Clara, landslides from both sides of the gulch in which the Loma Prieta sawmill was located completely buried the mill, entombing the nine men working there.

In San Francisco, which then had a population of about 425,000 and sported a score of steel-frame high-rise office buildings, structural damage was widespread but irregularly distributed. Brick walls, chimneys, and cemetery monuments tumbled throughout the city, but the most severe destruction was largely confined to low-lying areas of filled-in ground near San Francisco Bay. In some areas south of Market Street, the city's main thoroughfare, the intense shaking of soils with a shallow ground-water table caused the ground to fail and subside many feet. The Valencia Street Hotel sank into the earth, crushing and burying scores of its residents as the fourth floor dropped level with the street. Some persons were drowned when water from broken mains flooded the shattered lower floors. About 3000 persons died of earthquake-related injuries, and the death toll would have been far higher had the quake not occurred when most people were asleep in their beds rather than out on the streets in the path of falling walls.

The buildings that suffered the most severe damage were generally made of unreinforced masonry or had been erected on poorly consolidated soils, particularly landfills covering old marshes, mudflats, ponds or creek beds. The same kinds of filled-in areas experienced the heaviest damage in the 1989 Loma Prieta earthquake. Apart from edifices set on landfill, the most spectacular ruin was that of San Francisco's magnificent city hall. Recently completed at a then-phenomenal $7 million, this Greco-Roman structure resembled the Capitol in Washington, D. C., but with an even loftier dome supported by two tiers of stone columns. Exposing its pseudo-classical facade as the architectural sham it was, the earthquake stripped most of the stonework from the building's frame, leaving the great dome held aloft only by twisted steel girders. The city hall was already a symbol of municipal corruption; the earthquake made it a symbol of seismic devastation, a favorite subject of photographers.

But the earthquake, which caused far greater damage than civic leaders were later willing to admit, only began San Francisco's destruction. Almost as soon as the shaking stopped, fires broke out all over the downtown area. They started in broken gas mains, overturned stoves, severed electrical wiring, and houses where incautious residents built breakfast fires using quake-loosened chimneys. By noon a square mile of the city's heart was in flames, including much

The great earthquake of April 18, 1906, stripped San Francisco's city hall of its pseudo-marble facade, exposing the steel frame that held the dome aloft. This photo was taken after the fire had gutted the quake-ravaged structure. —California Division of Mines and Geology photo

of the working class residential area south of Market Street. About two o'clock that afternoon the fire swept north across Market to begin consuming the financial district and the hotels and fashionable shops around Union Square. By the next day Chinatown had vanished in smoke, and Nob Hill's mansions were aflame.

For three days and nights the conflagration raged unchecked, spreading over more than four square miles and leaving nearly 350,000 persons homeless. All the downtown fireproof buildings, which had demonstrated their superior design and construction by riding out the earthquake with only minor damage, were gutted in the holocaust. More than 28,000 buildings burned to the ground. Almost no water was available to fight the fire. The quake had not only broken most of the city's water mains, it had also severed the main supply lines connecting San Francisco with its reservoirs. Not until firemen, the military, and ordinary citizens impressed into service had dynamited almost the full length of Van Ness Avenue to create the most expensive firebreak in history was the fire's westward advance halted.

Today some city officials offer assurance that modern fire-fighting techniques and an improved quake-resistant water supply would prevent San Francisco from burning after another 8.3 magnitude earthquake. Many experts, including a former San Francisco fire chief, however, argue that San Francisco faces a fire hazard that is now greater than it was in 1906. The city's fire department was stretched to its limits following the Loma Prieta quake, centered nearly sixty miles away. A foretaste of what a larger earthquake will bring, the Marina District fire was a miniature replay of the 1906 disaster: ruptured gas mains sent flames leaping 300 feet into the air, consuming a dozen or more buildings; quake-broken water mains made water unavailable when most needed, forcing fire crews to pump seawater from San Francisco Bay many blocks distant. Without the fireboat Phoenix supplying water offshore, the Marina blaze could have spread catastrophically. A 1987 insurance report suggests that a minimum of seventy-nine major fires would break out in different parts of the city following a large earthquake, thirty-nine of which could require more equipment and personnel than the fire department, much smaller than it was in 1906, now possesses. Considering that the 6.4 magnitude San Fernando quake of 1971 started 109 fires, the insurance report probably underestimates San Francisco's potential hazard.

San Francisco's business district after the 1906 earthquake and fire. The largest urban conflagration in American history, the fire destroyed more than 28,000 buildings and left up to 350,000 persons homeless. The slim domed tower at the left of center is the Claus Spreckles building, at nineteen stories then the tallest structure west of Chicago. —California Division of Mines and Geology photo

Despite the certainty of future large earthquakes, some as strong as those of 1857 and 1906, Californians need not worry about their state's falling into the ocean. Although some journalists have exploited fears that an imminent cataclysm would cause California west of the San Andreas fault to sink beneath the Pacific, scientists completely dismiss this alarmist view. Throughout the tens of millions of years that the oceanic plate has crept and lurched northward with respect to the continental plate, movement along the San Andreas Fault has been almost entirely horizontal, not vertical. In 1906 maximum horizontal displacement was about twenty-two feet, with relatively little vertical change. Over the next few million years, the moving crustal block will carry Los Angeles north to the San Francisco suburbs—an eventuality that many northern Californians may regard as a catastrophe worse than that of 1906, but not really equivalent to having the state suddenly disappear beneath the Pacific.

Owens Valley, East of the Sierra, 1872

Although the San Andreas fault produces our nation's most celebrated earthquakes, the east-central part of California experienced an event that possibly equalled them in size. On March 26, 1872, the Owens Valley fault that parallels the steep eastern face of the Sierra Nevada shifted violently, cutting a deep furrow in the ground surface for at least 100 miles from Haiwee Reservoir south of Olancha to Big Pine. In this case vertical movement was dramatic, particularly between Lone Pine and Independence, where relative displacement on opposite sides of the fault totalled twenty-three feet. The Owens Valley fault is related to the Sierra fault system, where colossal fractured blocks of crustal granite have been raised during millions of years to form the Sierra peaks. As the mountains rise, the crustal block underlying Owens Valley to the east sinks, generating innumerable earthquakes in the process.

Famed naturalist John Muir was camping in Yosemite Valley the night the 1872 earthquake arrived. Exhilarated by the immensity of the event, he vividly described the clash and roar of huge granite boulders avalanching down the Yosemite canyon walls, striking sparks and electrical flashes as they crashed in midair before thundering to earth.

Near the Owens Valley epicenter, destruction was nearly total. At the town of Lone Pine, which then boasted between 250 and 300 inhabitants, fifty-two of its fifty-nine adobe houses were leveled, killing twenty-three people. The earth heaved so violently that fish were thrown out of the Owens River onto the riverbanks, while a huge wave developed on Owens Lake. Eyewitnesses reported that water withdrew from the shore, standing in an apparently vertical wall out in the lake, before it returned with unexpected gentleness to overflow the shoreline. Considering the size of the earthquake, estimated to have been about 8.0 on the Richter scale, the death count of sixty was low because very few people had settled in this arid terrain. This barren and wind-swept part of California still has few towns and few people, but a repetition of the 1872 event could endanger thousands of hikers, campers, skiers or other visitors to Yosemite, Mammoth Lakes, or others parts of the Sierra.

Volcanic Earthquakes

Earthquakes have been common lately north of Owens Valley, in the Mono and Mammoth lakes region. Beginning in May, 1980, when a series of four earthquakes registering a Richter magnitude of about 6.0 occurred during one forty-eight hour period, the community of Mammoth Lakes has shaken frequently. Unlike the 1872 shock, the current swarms of earthquakes are thought to be caused by a large mass of molten rock sporadically rising toward the surface. In 1989 another series of small quakes centered beneath Mammoth Mountain, an old volcano near Mammoth Lakes. Earth scientists believe that these earthquakes may presage a volcanic eruption somewhere in the area, perhaps within the next few decades (see Chapter 16).

As if California's many recognized faultlines were not enough to worry about, geologists warn that numerous concealed faults may also be potentially dangerous. Before it caused the 1987 quake that severely damaged the town of Whittier, near Los Angeles, the hidden Whittier-Narrows fault was not known to exist. Many other buried faults may lie beneath the Los Angeles basin, as well as other parts of California.

California has not experienced a great earthquake since 1906, although it was struck by three catastrophic quakes during the half century preceding that date. The major quake of October, 1989, which cost sixty-seven lives and caused billions of dollars in damage nearly sixty miles from its epicenter, is only an intimation of what will happen on a far greater scale when a magnitude 8.0 earthquake strikes our most populous state. Geological processes are inexorable, and sections of the San Andreas and other active faults that are now locked may break free at any time, generating shocks that will shatter both buildings and lives.

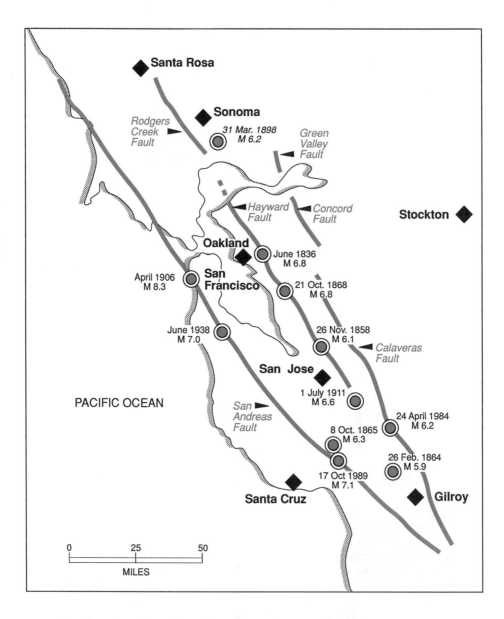

San Francisco sits midway between two major earthquake faults, both of which are capable of devastating the entire San Francisco Bay region. Note that the Hayward Fault, running through the East Bay, produced major shocks in 1836 and 1868, while the San Andreas was responsible for the catastrophic 1906 earthquake that shattered cities along 280 miles of the northern California coast.

Marking the boundary between the Pacific plate and the North American plate, the San Andreas Fault cuts through the Corrizo Plain in central California like a huge furrow ploughed by giants. The dark line left of the fault is vegetation growing along a fence that runs parallel to the fault, about fifty miles north of Santa Barbara.

—California Bureau of Mines and Geology photo

Chapter 7

WHEN THE "BIG ONE" HITS CALIFORNIA

The 1989 earthquake that toppled buildings, bridges, and freeways from Santa Cruz to San Francisco was not the anticipated "Big One." The Loma Prieta temblor scored a 7.1 magnitude on the Richter scale; the Big One will be a repeat of the 8.3 shock that decimated the northern California coast in 1906, an event more than thirty times stronger than the jolt of '89. The inferno in the Marina District and the carnage on the Cypress Structure in west Oakland were only local neighborhood dress rehearsals for the great devastation that will sweep through much larger sections of California in the near future.

Eighty-five percent of California's population is concentrated in two large urban areas that live in double jeopardy. San Francisco sits midway between two major earthquake faults, the northern San Andreas and the Hayward fault. The Los Angeles metropolitan area sprawls between the southern San Andreas and the Newport-Inglewood fault. All four active fault zones are capable of triggering

47

Car smashed by falling bricks from a masonry structure near Interstate 880 in west Oakland, October 17, 1989. Debris cascading from the collapsed upper stories of a brick building crushed cars in San Francisco, killing five people. —California Division of Mines and Geology, David R. Montgomery photo

powerful earthquakes that will kill many thousands of people and cause tens of billions of dollars in property damage.

The San Andreas fault slipped about five feet to generate the Loma Prieta quake that killed sixty-seven people and caused property damage of up to $10 billion. Recent studies by the California Division of Mines and Geology suggest that a repeat of the 1906 San Francisco earthquake, in which the San Andreas moved as much as twenty-two feet, could kill 11,000 people and critically injure another 44,000. A 7.5 Richter-magnitude quake on the Hayward fault, which runs through the East Bay region, is expected to cause about 7,000 deaths. In the greater Los Angeles area, twice that number would die during a magnitude 8.3 shock on the southern San Andreas. A magnitude 7.0 temblor on the Newport-Inglewood fault, which cuts through the densely populated Long Beach and Compton areas, may claim as many as 23,000 lives. Many hundreds of thousands, perhaps millions, would be left homeless. All of these figures are conservative estimates and do not take into account the effects of large fires that typically follow great earthquakes.

Nor do government statistics convey the horror, shock, and personal loss that millions of Californians will suffer. Survivors may experience a psychological and social dislocation as severe as the shaking that will render many of their homes and businesses uninhabitable.

Whichever urban center the Big One hits, the suddenly heaving earth will transform many areas into death traps. Thirty to forty times more powerful than the 1989 event, the great earthquake will shake harder, longer, and over a much larger region. Tall office buildings will rock and sway violently, hurling desks, steel filing cabinets, and light fixtures at their inhabitants. Razor-sharp shards, guillotines of glass, from countless shattered windows of high rise structures will rain down on pedestrians and cars below. Cornices, storefronts, and some entire buildings will crash into crowded streets, choking them with debris. As in '89, many older freeway overpasses and bridges will collapse, crushing hundreds of cars and their occupants. Endless lines of stranded vehicles will effectively halt traffic on most highways, while landslides will block others. Broken gas mains and oil and gas refineries will erupt in flames. Communications and electrical and other power sources will fail on a massive scale, leaving survivors isolated in near-total darkness after nightfall. Damage that was largely confined to small towns near the epicenter of the Loma Prieta earthquake and to isolated neighborhoods in the Bay Area will be widespread along a broad swath of the San Andreas faultline.

Northern California

Despite its glamor and sophistication, San Francisco lives metaphorically in the calm eye of a hurricane. After being largely destroyed by the earthquake and fire of 1906 and jolted into a renewed awareness of its vulnerability in 1989, the city more or less serenely awaits another cataclysm that may occur within the next few decades. Downtown San Francisco occupies a particularly hazardous position. Located almost equidistant from the San Andreas and Hayward faults, the business district, with its forest of 500 steel and glass skyscrapers, is built partly on water-saturated sediment of a kind that typically liquefies during the violent shaking of an earthquake. Subsurface liquefaction causes widespread ground failure, slumping and subsidence that buckle and crack streets, sever gas and water mains, and wreck foundations, triggering a building's collapse. The collapse of a single large office building during working hours may kill as many as 1,000 people, with relatively few escaping with only minor injuries.

49

*Ground liquefaction and lateral spreading in San Francisco's Marina
District buckled sidewalks and wrecked many foundations. The badly
damaged building in the center was later demolished.*
—U.S. Geological Survey, M. Bennett photo

In 1906 and 1989 much of the worst structural damage occurred on
filled-in land, areas where earth had been dumped to create new
building sites along the shore of San Francisco Bay. The entire San
Francisco waterfront, including the eastern part of the city's financial
district, as well as most areas bordering the East Bay, lie on water-
logged landfill that will suffer maximum shaking intensities during
a great earthquake.

How dangerous erecting high-rise edifices on top of such unstable
fills can be was graphically demonstrated in the Mexico City earthquake of 1985, which killed approximately 10,000 persons. Almost all
the hundreds of modern steel-frame structures that collapsed in
Mexico City were on an ancient lake bed with a high water table.
Because similar kinds of mud underlie thousands of structures
fringing San Francisco Bay, applying the lesson of Mexico City to
stricter building codes in this area may be a matter of life and death
to thousands of persons.

California's Seismic Safety Commission estimates that San Fran-
cisco has at least 2,000 buildings likely to collapse in a strong quake.
Built of unreinforced masonry, many of these unsafe buildings are in
Chinatown and south of Market Street. Most are low-rent housing. To

make them quake-resistant could cost up to $30.00 per square foot, an expense owners are unlikely to undertake without raising rents and adding to the city's ever-growing crowd of homeless persons.

Eager to rebuild after the 1906 fire, San Francisco officials lowered building code standards so that for several decades thousands of structures were hastily thrown up without meeting pre-1906 earthquake and fire resistant criteria. Many will come down even more quickly during the Big One. Some have since been structurally reinforced or replaced with better designed structures. Although most well-constructed wood-frame buildings fare better in an earthquake than unreinforced masonry structures, in 1906 hundreds of poorly built wooden business and apartment buildings disintegrated into kindling wood, killing several hundred people. Today many neighborhoods in San Francisco are filled with similar structures that may fail catastrophically.

Whether the Big One originates on the San Andreas or the Hayward fault, San Francisco will be hard hit. Another magnitude 8.0 or higher on the same segment of the San Andreas that broke in 1906 will also cause severe damage along the entire northern California coast from Monterey and San Jose to Santa Rosa and Point Arena. In 1906 fault movement ruptured the ground surface for a distance of 280 miles, with a maximum displacement of twenty-two feet in Marin County. In 1906 about 500,000 people lived in the affected area, compared to over 6 million today. Because of the vast population increase, the deadly effects of future earthquakes will be commensurately greater.

Because the San Andreas fault roughly parallels the main highways that traverse the San Francisco peninsula connecting the city to other urban and suburban centers, San Francisco is likely temporarily to be cut off almost entirely from ground-level communication. Massive land subsidence, surface cracking, landslides, and the collapse of numerous bridges and overpasses will block virtually every arterial, both north and south of the city, including U. S. Highway 101, Route 1, El Camino Real, and Route 17. The 1989 quake not only toppled a large section of the Interstate 880 freeway in the East Bay, it also seriously damaged parts of Interstate 280 and the Embarcadero freeway in San Francisco, closing them indefinitely. Route 17 connecting Santa Cruz and San Jose was closed for over a month. After the expected great earthquake, many Bay Area freeways will remain impassable for weeks to months.

The quake of '89 damaged several buildings at San Francisco International Airport, closing it for several days. A magnitude 8.3

Close-up of sand boils and lateral spreading crack on taxi-way near the northwest end of the main runway of Oakland International Airport. Severe shaking of unconsolidated landfill generated soil liquefaction and ground failure that damaged about one third of the runway. When the anticipated 8.3 shock hits northern California, it is expected to render airports throughout the entire San Francisco Bay area inoperable.
—U.S. Geological Survey, T. Holzer photo

shock will damage all airports in the Bay Area sufficiently to prevent large cargo planes from landing. The closest usable airfields may be Buchanan near Concord and Travis near Vacaville. Only helicopter transport can be counted upon to bring medical, food, and other supplies into the stricken area from the outside. Emergency relief workers will have to explore ways of transporting supplies to the city by water. Horizontal movement of the fault is not expected to generate tsunamis and most Bay Area piers and docks are solidly anchored on piling that is expected to withstand the earthquake.

While a large magnitude quake on the San Andreas is likely to cause considerable damage to older structures, including bridges and freeway overpasses built on landfill in the East Bay area, the worst damage may be confined to a relatively narrow corridor along the Pacific coast. The Stanford campus in Palo Alto suffered heavy damage in 1906 and 1989, but the University of California at Berkeley, only twenty miles east of the San Andreas fault, escaped largely unharmed.

The Hayward Fault

Oakland, Berkeley, and other populous East Bay cities will suffer most severely if the coming quake occurs on the Hayward fault. Running parallel to the San Andreas for sixty-two miles through the East Bay region, the Hayward fault can inflict extensive damage on both sides of San Francisco Bay. In 1836 and 1868, the Hayward generated strong quakes that toppled buildings from Oakland to San Francisco. Many geologists believe that a 7.5 magnitude shock on the Hayward fault could claim as high or higher a toll in lives and property as a larger event on the San Andreas.

Horizontal movement averaging about five feet along the full length of the Hayward fault, the maximum displacement that earth scientists think probable, would cause death, destruction, and mass confusion throughout the Bay Area. Although a recent study suggests that major highways, bridges, sources of communication, power, light, and other utility lifelines would probably be cut off for a maximum of seventy-two hours, the 1989 earthquake demonstrated that large areas can be without power or drinkable water for much longer. Streets blocked by rubble, and damaged highways and bridges may effectively isolate communities for days. At a time when they are most urgently needed, indispensable services suddenly will be un-available to millions of people.

This bridge in Santa Cruz suffered extensive damage in the Loma Prieta quake. Post-quake road and bridge closures may isolate communities for days, cutting off access to emergency services and supplies. —Annie Kappl photo

The 7.5 magnitude quake will produce strong shaking from near Petaluma and Napa in the north Bay to south of San Jose, causing significant damage in ordinary well-built structures and severe damage in poorly constructed buildings. Earth scientists believe that the greatest damage will occur in a zone about five miles wide lying generally west of the Hayward fault. This high-risk zone encompasses nearly all the intensely developed area of the East Bay from San Pablo southeast to the eastern half of San Jose. Even specially designed buildings are expected to suffer considerable damage. Well-built structures will partially collapse, while countless houses and other buildings will be thrown off their foundations.

Before October 17, 1989, when a large section of the San Francisco-Oakland Bay Bridge collapsed, killing one driver, most engineers expected the great suspension bridge to survive a major earthquake intact. State projections of quake damage to the Golden Gate and Bay bridges anticipated that their approaches would partially collapse but

A collapse that engineers predicted would not happen. Built in 1936, the San Francisco-Oakland Bay Bridge was designed to resist strong earthquakes. Although centered almost sixty miles away, the 1989 Loma Prieta quake produced a violent shaking that severed the six-inch long steel bolts anchoring the bridge's upper deck. Manufactured to withstand a force of 500,000 pounds, the bolts were subjected to a force four times more powerful. —California Department of Transportation, Bob Colin photo

The demolition of the Veterans Hospital by the February 9, 1971, San Fernando earthquake on the southern San Andreas fault foreshadows what may happen to numerous hospitals and other essential structures built near the Hayward fault which runs through the populous east San Francisco Bay area. A 7.5 magnitude earthquake on the Hayward fault may claim a higher toll in lives lost than a larger quake on the San Andreas. —

U.S. Geological Survey, Tim Hall photo

that their main spans would stand unscathed. The Bay Bridge was reopened a month after the earthquake, but public confidence in the structure's ability to resist a significantly larger quake at a much closer epicenter was understandably shaken.

The number of persons killed by a 7.5 quake depends upon the time of day that the event takes place. The least dangerous time for a major quake would be the early morning hours when most people are asleep. An earthquake occurring at 2:30 a.m. might cause about 1,500 deaths and 50,000 injuries; more than 4,500 of the injured would require hospitalization. By contrast, an earthquake striking at 2:30 p.m., when many people are working, might kill 7,000 and hospitalize 13,200, with another 132,000 suffering lesser injuries.

One of the most disturbing factors in planning for future Bay area earthquakes is that eight of the twenty-six general acute care hospitals in Alameda and Contra Costa counties are located within one mile of the Hayward fault. It is a grim joke that one can trace the Hayward fault on a map by drawing lines that connect hospital sites. Most of these hospital buildings, including a large U. S. military facility, do not meet the construction safety standards that California imposed after several hospitals collapsed in the 1971 San Fernando earthquake. In addition to the loss of life from damage to the hospitals themselves, at least thirty-five percent of the 6,200 beds available in these crucial facilities may be rendered partly or largely unusable.

The Hayward fault also slices through the University of California campus at Berkeley, running directly beneath the U. C. Memorial Stadium. It also passes beneath or near such essential nerve centers as police and fire stations and utility services throughout the East Bay. All the major freeway routes to the East Bay from the east and south either cross the fault or are vulnerable to closure from landslides, ground failure, or other disruption. Some of California's most heavily traveled highways, including Interstate 80 at San Pablo, Interstate 580 in East Oakland, and Interstate 680 at Fremont could suddenly offset as much as five feet, horizontally shifting one lane half into line with the next. Ground subsidence and liquefaction could make long sections of Interstate 80 and 880, already partly destroyed by the 1989 earthquake, and Route 17 from Richmond to San Jose largely impassable.

Damage to transportation corridors, including highways, railways, and airports, could seriously hinder emergency services. It will also prolong the time necessary to restore damaged electrical power, telephone, and gas lines. Engineers predict that Oakland and other East Bay cities will be totally without power for at least twenty-four to forty-eight hours. In San Francisco, fully half the residents will have no electrical power during the first twenty-four hours and about twenty-five percent will experience power outages for an additional day. In the actual event, power outages will probably be even more extensive and last significantly longer.

The disruption of water mains, aqueducts and distribution systems that cross the Hayward fault will create acute water shortages in the East Bay, in some areas for several days. Lack of potable water, plus contamination of ground water and water from broken sewer lines, may produce serious threats to public health.

Fault movement will also cause thousands of breaks in gas mains, valves, and service connections. The six principal Bay Area petroleum refineries, located at or near the margins of San Pablo and Suisun Bays, are likely to suffer heavy damage. All major pipelines carrying petroleum fuels to the region cross the Hayward fault either at San Pablo or Fremont, where ground movement will damage or rupture them. With most major highways blocked, water supplies curtailed, and gas and oil pipelines broken over a wide area, fires will be inevitable. Some may burn out of control for days.

A 1987 insurance risk report estimates that a large earthquake will ignite so many fires in San Francisco that the fire department, smaller than it was in 1906, will be unable to control them. A minimum of 142 fire engines will be needed to handle the crisis, four times the number of engines the city fire department had in 1989. With water unavailable in many neighborhoods, the fires are expected to destroy between 22,000 and 48,000 buildings, creating a holocaust larger than the 1906 conflagration, which is still the worst urban fire in world history.

The Southern San Andreas Fault

Many earth scientists expect a magnitude 8-plus quake to devastate southern California within the next thirty years. During the late 1970s, Kerry Sieh, a geologist at the California Institute of Technology in Pasadena, explored the sequence of major movements along the southern San Andreas fault. By digging trenches across the faultline, Sieh was able to study the number and size of fault displacements, movements that had carried rock formations on one side of the fault away from their original positions. He found that at least twelve large earthquakes had occurred during the last 2,000 years, averaging one about every 140 years. The year 1987 marked the 130th anniversary of the last great earthquake on that section of the fault, the Fort Tejon event of 1857.

The earth's creeping skin does not operate on a precise statistical timetable, and no reputable scientist would cite 1997 as the date for the Big One in southern California. In fact, recent studies indicate that sporadic movement along the southern San Andreas has been more complex than originally inferred. Great events are random, too linked to the irregular pulse of chaos to be predictable.

If a future earthquake the same size as the 1857 Fort Tejon quake, a magnitude 8.3, originates on the same south-central segment of the San Andreas fault, intense shaking will occur across a broad belt of southern California. Although the southern San Andreas fault is not as close to urban centers as it is in the San Francisco Bay area, damage will be widespread and costly. If the earthquake strikes in mid-afternoon, when most people are in crowded business and shopping areas, it may kill as many as 14,000 people and seriously injure 55,000 more. Tens of thousands of people will suffer some degree of injury or disability. The enormous number of injured and homeless people and the failure of many electrical power, telephone, and utility sources will overwhelm the state or local government's ability to cope with public needs.

Areas in a broad zone along the faultline will experience intense shaking and commensurate damage to office buildings, homes, highways, and other structures. Areas north of the fault underlain by alluvial materials, such as the Antelope Valley, will also shake violently. Sites underlain by mountain bedrock will suffer less damage.

If the San Andreas fault rupture extends farther southeast than it did during the Fort Tejon earthquake, the San Bernardino area will also experience maximum shaking intensities. Although shock intensities will gradually diminish with increasing distances from the fault, densely populated urban areas with a high groundwater level

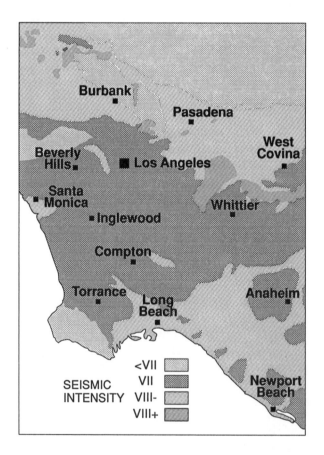

A magnitude 8 earthquake on the southern San Andreas fault would cause severe shaking and damage to structures throughout a much larger part of southern California than a 7.0 event on the Newport-Inglewood fault, but the intensity of shaking would be less powerful in the Los Angeles area, which is located farther from the San Andreas. Note that many communities surrounding Los Angeles, including Long Beach, Burbank, and Pasadena, would experience an estimated Mercalli intensity of about VIII. —After Toppozada and others, 1989

Jefferson Junior High School in Long Beach, California, suffered total collapse during the 1933 earthquake. Damage to public schools was so extensive, the California legislature passed the Field Act mandating earthquake-resistant building codes for all state schools.
—California Bureau of Mines and Geology, M. Heinvel photo

or underlain by soils susceptible to liquefaction can suffer major damage. These include the Long Beach and Huntington Beach areas, the Santa Clara Valley, the Ventura-Oxnard areas and much of the San Fernando Valley.

Unlike San Francisco after a magnitude 8.3 shock, the Los Angeles region probably will not be cut off from ground level communication with the outside world. Landslides will block highways locally, including mountainous sections of Interstates 5 and 15, but U. S. Highway 101 along the coast should remain open as should highways from San Diego.

The Newport-Inglewood Fault

Severe as the effects of an 8.3 shock on the southern San Andreas will be for the Los Angeles region, the consequences of a magnitude 7.0 on the Newport-Inglewood fault will be far greater because that fault traverses the highly populated Los Angeles Basin. The fault extends from the ocean floor south of Laguna Beach and runs approximately forty-five miles northwestward to Beverly Hills. The earthquake of December 8, 1812, which destroyed the bell tower at Mission San Juan Capistrano and killed forty Indians attending mass in the church, probably originated on the Newport-Inglewood fault. An estimated magnitude 7, the 1812 shock also extensively damaged the missions at San Gabriel and San Fernando.

In terms of lives lost, the Newport-Inglewood fault produced the second worst quake in California history, the Long Beach-Compton earthquake of March 10, 1933. About one hundred twenty persons perished when a subsurface rupture occurred at a depth of six miles just off the coast of Newport Beach. Damage from the 6.3 temblor was so extensive, particularly to schools and other public buildings, that the California legislature passed the Field and Riley acts to regulate, respectively, the construction of public schools and of dwellings built to hold more than two families.

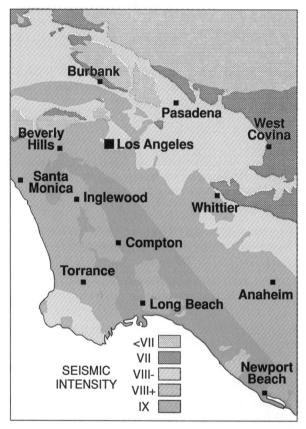

Red denotes the region of maximum shaking from a projected 7.0 earthquake on the Newport-Inglewood fault. Note that almost the entire Los Angeles-Long Beach-Newport area could experience Modified Mercalli intensities greater than VIII, strong enough to cause severe damage throughout much of the Los Angeles basin. The 1989 Loma Prieta earthquake produced shaking intensities of up to IX in scattered locales around San Francisco Bay, devastating the Marina district and parts of west Oakland.

The scenario for a magnitude 7.0 earthquake on the Newport-Inglewood fault postulates a subterranean movement averaging about three feet for forty-five miles along the fault. For approximately twenty-five seconds the ground will shake violently over a wide area, almost twice as long as the San Francisco Bay Area trembled in 1989. Shaking will be most intense in soft ground along a zone five miles wide for the entire length of the fault, including the areas of Long Beach, Compton, and Newport Beach. Although a magnitude 8.0 shock on the San Andreas fault will cause damage throughout a larger area, including parts of San Bernardino, Kern, and Ventura counties, the magnitude 7.0 event on the Newport-Inglewood fault will have a greater impact on the densely populated Los Angeles metropolitan area.

Fault movement and ground deformation along the Newport-Inglewood fault zone will damage many more highways, utilities, essential service buildings, and industrial facilities in the Los Angeles metropolitan area than will a much larger movement of the San Andreas fault. North-south transportation corridors, including many of southern California's most heavily traveled freeways, may be too damaged to use. Stalled and abandoned cars on offramps and over-passes will further impede transportation. More than 130 miles of state highways and over 350 state bridges in the Los Angeles area are vulnerable to critical damage. Interstate Highway 5, which lies east of the zone of maximum shaking in a Newport-Inglewood event, may provide an alternative north-south transportation route.

The region's five principal airports, including Los Angeles International, which services about 250,000 people daily, will suffer varying degrees of damage. Damage to access freeways, control towers, fuel tanks, and other structures will severely limit use of airports that are still operating.

Broken water, sewage, and petroleum fuel lines will create major problems of environmental pollution. The principal waste water lines into the Carson treatment plant will suffer heavy damage at the fault crossing between Compton and Long Beach. Damage and lack of fresh water for treatment and of electrical power for pumping will dump raw sewage into soils, channels, and streets, contaminating the ground water and the coastline. In Los Angeles Harbor, broken oil pipelines and storage facilities will discharge oil into the channel. Similar damage to petroleum tanks and piping in Seal Beach will cause oil spills into Alamitos Bay. Damage to petrochemical plants may permit widespread release of toxic emissions.

The near-collapse of this two-story building in the 1983 Coalinga earthquake is typical of damage inflicted on unreinforced masonry structures. Although the Coalinga quake was of only moderate size, it showed that the intensity of shaking at a given location near the epicenter can be as strong as that produced by much larger earthquakes.
—California Bureau of Mines and Geology photo

Explosions and fires are likely at oil and gas storage facilities in Los Angeles and Long Beach harbors. Another major conflagration may break out on Mormon Island. Thousands of shattered natural gas mains and petroleum fuel lines, both along the fault zone and at connections to individual houses and other buildings, will ignite fires throughout the area. Some blazes may rage for days, scattered over a wide area, from port facilities to suburban housing tracts. The many fires could pose the greatest challenge to American fire crews since the San Francisco holocaust of 1906.

With water supplies and electrical power curtailed and with most communication lines cut, for several days after the quake coordination of emergency relief to the tens of thousands of injured persons requiring immediate help will be painfully difficult. Hundreds of thousands of homeless persons seeking food and shelter will further strain sources of relief. The 1989 quake stretched the San Francisco Bay Area's fire, police, and other emergency crews to their limit. A great earthquake affecting a much larger region, followed by uncontrolled fires, would far exceed state and civic agencies' ability to cope. Other social problems stemming from the earthquake are difficult to assess. In the crisis following a great natural disaster, most survivors behave responsibly, even heroically, as did the majority of San FrancGGGiscans in 1906 and 1989. Some segments of the population may panic while others loot.

Built along the moving edges of continent and sea floor, both San Francisco and Los Angeles face a seismically violent future. San Francisco endures a double threat from the San Andreas and Hayward faults, both of which can produce earthquakes strong enough to devastate the Bay Area. Although the Los Angeles metropolitan area will suffer widespread damage from a repeat of the 1857 quake on the San Andreas fault, it is likely that a magnitude 7.0 shock on the Newport-Inglewood fault will prove significantly more destructive. A moderate quake on the Newport Inglewood fault shattered Long Beach, Compton, and other near-by settlements in 1933. The next major shake may equal that of 1812, which rattled an area now inhabited by twelve million persons.

Registering a Moment magnitude of 9.2 ,the 1964 Prince William Sound earthquake devastated many of Alaska's cities. Note the effects of ground subsidence in downtown Anchorage. —California Division of Mines and Geology photo

Chapter 8

PLATES IN COLLISION: ALASKA'S "SUPERQUAKES"

Unlike the San Francisco earthquake of 1906, the vastly more powerful earthquake that devastated south-central Alaska in 1964 has not become part of American folklore. Perhaps because the forty-ninth state seems so physically remote to most Americans, the significance of this largest quake ever recorded in North America has not entered the national consciousness.

It struck on the Friday before Easter, March 27, 1964, at an epicenter in Prince William Sound, about eighty miles east of Anchorage, home to approximately half of Alaska's total population. The earthquake originated about 12.5 miles beneath the eastern shore of Unawik Inlet in northern Prince William Sound. It was triggered by massive fracturing of the sinking edge of the northeastern Pacific Ocean plate where it descends beneath the rocky slab carrying the North American continent

This slow-motion collision between the southeastward moving ocean floor and the northwestward drift of the continent, at the rate of about four or five inches per year, has generated many of the world's greatest earthquakes. Between 1899, when instrumental seismic recording began, and 1961, seven Alaska earthquakes scored 8.0 or higher on the Richter scale. More than sixty equalled or exceeded magnitude 7.0. About seven percent of the seismic energy released annually on earth originates in the unusually active subduction zone along Alaska's coast. Because it released far more energy than could be measured by a Richter magnitude 8.0, the 1964 Prince William Sound earthquake is assigned a Moment Magnitude of 9.2, making it second only to the 9.5 Chile earthquake of 1960, the strongest yet recorded.

The 1964 Alaska earthquake opened surface fissures wide and deep enough for adult men to hide in along the Seward-Anchorage Highway near Portage. Subsidence and lateral ground displacement were responsible for surface cracking, which was common throughout the affected area.
—U.S. Geological Survey, R. Kachadoorian photo

Striking shortly after 5:36 in the afternoon of a typically overcast day in early spring, the bone-jarring main shock lasted for a nightmarish *four minutes*, considerably longer than the duration of most great earthquakes. The violence of the ground motion was also remarkable, causing the surface to pitch and roll like a ship at sea and throwing thousands of persons to the ground. The giant quake was felt over all of Alaska and large portions of western Yukon Territory and British Columbia, an area totaling approximately one million square miles.

Within seconds huge sections of the earth's surface rose or sank many feet. Portions of former seafloor along the coast became dry land. On Montague Island maximum uplift was thirty-eight feet, and at Cordova, seventy-five miles from the epicenter, docks rose about eleven feet. Simultaneously, vast tracks of land and sea-bottom subsided, including the northern and western parts of Prince William Sound, the western segment of the Chugach Mountains and portions of the lowlands north of the mountains, most of Kenai Peninsula, and almost all of the Kodiak Island group. Altogether, over 100,000 square miles of rugged mountains, ice-fields, fjords, glaciers, and seashore moved vertically, more surface distortion than that associated with any other known earthquake.

The violent buckling of the ocean floor, accompanied by massive submarine landslides, generated tsunamis that scoured much of the Alaska coast. At Valdez, a busy port built on unstable glacial deposits, an underwater delta slid into the sea, carrying with it the town docks on which thirty persons had been waiting for a merchant ship. All were borne to their deaths in watery chaos. Kodiak Island, Seward, and Homer, a picturesque fishing village located on a low spit at the tip of Kenai Peninsula, were also swept by deadly tsunamis. All these towns experienced severe damage, but many smaller coastal settlements suffered even heavier property losses in proportion to their size. Tsunamis generated by the Alaska earthquake proved lethal as far south as Crescent City, California, where twelve persons drowned when waves twenty feet high rolled inland.

Anchorage, with a population of about 100,000, suffered spectacular damage from both vertical and horizontal ground movement. Lateral movement along layers of silt and clay beneath the city's surface caused widespread and destructive fissuring as the ground broke irregularly into yawning cracks and deep pits. A fashionable residential neighborhood on Turnagain Bluff lost about seventy-five expensive homes when 200 acres abruptly slumped 500 feet downhill toward the sea. In downtown Anchorage, many commercial buildings were demolished when large blocks of earth along L Street, Fourth

Huge sea waves, tsunamis, generated by the March 27, 1964 Alaska earthquake swept ocean-going vessels well above the high tide-line into the heart of Kodiak, Alaska. —U.S. Geological Survey, R. Kachadoorian photo

Avenue, and Government Hill dropped five to fifteen feet. Nearly every major building experienced significant damage, in many cases total collapse. Large concrete slabs facing the new Penneys' Department Store shook loose and slammed into the street, crushing pedestrians on the sidewalk. A nearly completed apartment house six stories high telescoped into a pile of rubble. With 114 persons dead, $350 million worth of property destroyed, and thousands of Alaskans suddenly without homes or jobs, only prompt and relatively well-coordinated relief efforts prevented the 1964 earthquake from becoming a lasting social and economic catastrophe.

The Future

Steady commercial development and population gains increase Alaska's vulnerability to natural disasters, as well as to man-made debacles such as the 1989 Exxon oil spill near Valdez. With every passing decade, the earthquake and volcanic hazards posed by Alaska's intensely active subduction zone become a real threat to more people.

Rapid plate movement along southern coastal Alaska and the Aleutian Islands creates a long seismic danger zone that affects many different parts of a vast region. The Alaskan subduction zone consists of several segments which, on average, produce a great earthquake

Scores of expensive houses in Anchorage's Turnagain Heights were demolished in the massive landslide triggered by the March 27, 1964 earthquake. The violent shaking caused liquefaction of the clayey soil and subsequent massive ground failure. —U.S. Geological Survey, U.S. Army photo

The strongest earthquake ever recorded in North America devastated south-central Alaska on March 27, 1964. Violent shaking caused liquefaction of water-logged clay in the soil, resulting in widespread ground failure and subsidence. This photo shows Anchorage's Fourth Avenue, one side of which dropped six feet below the pre-quake street level. —California Bureau of Mines and Geology

comparable to that of 1964 every fifty to one hundred years. Less powerful quakes are equally destructive when they occur near population centers.

If a segment of the subduction zone has not produced a large earthquake in many years, earth scientists worry. Part of the descending plate may be temporarily snagged, slowly building up the energy to break free and trigger a high magnitude earthquake. Three quiet segments of the North America-Pacific plate boundary are now under suspicion. The first seismic gap, without recent large movements, is the Yakataga area between Icy Bay and Kayak Island. A decade ago the Unites States Geological Survey issued a Notice of Potential Hazard, warning that strain accumulating in the Yakataga Gap may be released soon in one or more earthquakes of magnitude 8.0 or larger. The longer strain accumulates, the more powerful the ensuing earthquake.

The Shumagin Gap lies farther west near the tip of the Alaska Peninsula. This area was the site of large shocks in 1788, 1847 and, possibly, 1903. Previous large quakes in the Shumagin area created

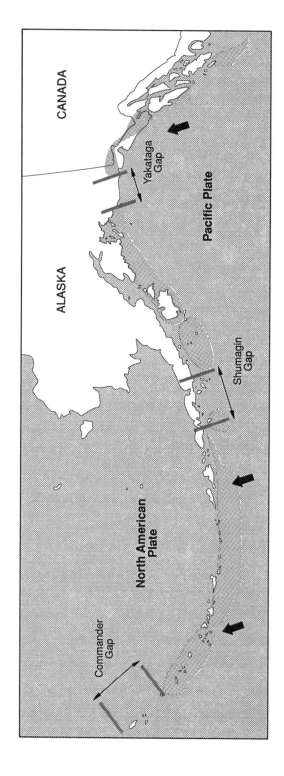

Subduction of the northern Pacific plate along the Alaska coast generates some of the world's largest earthquakes, including the M 9.2 cataclysm of 1964. Three sections of the subduction zone that have not produced large earthquakes in recent decades — "gaps" in the seismic process (red segments) — may produce earthquakes as destructive as the 1964 event in the near future. Because Alaska's earthquakes commonly trigger large tsunamis that sweep across the Pacific at high speeds, future quakes are likely to inflict serious damage on coastal settlements from Hawaii to California and the Pacific Northwest.

The 1964 Alaska earthquake caused widespread ground failure and subsidence, demonstrated here in the Government Hill Slide, Anchorage. —U.S. Geological Survey, Bonilla photo

large tsunamis, with wave crests reaching as high as one hundred feet. Future earthquakes are expected to generate equally destructive tsunamis that will threaten shipping and settlements along coastal areas hundreds of miles from the epicenter.

The Commander Gap lies at the extreme northwestern end of the Aleutian Islands. This remote section of the Alaskan subduction zone has not produced a major earthquake since 1858. When the fault finally moves, it will undoubtedly trigger large tsunamis. In 1946 a magnitude 7.2 earthquake south of Unimak Island in the Aleutian chain sent waves twenty to thirty-two feet high over the breakwater in Hilo, Hawaii, 2,300 miles from the epicenter. The tsunamis destroyed Hilo's waterfront, flooded the downtown area, and killed 159 people. Waves over one hundred feet high demolished a lighthouse at Scotch Cap on Unimak Island, crushing the five men inside.

Between 1946 and 1964, more than 200 people died as Alaskan earthquakes sent giant sea waves sweeping across the Pacific, claiming victims in areas as far apart as Valdez, Hawaii, and northern California. Because of their high frequency, enormous power, and wave-generating potential, Alaska's superquakes affect people and property at phenomenal distances from their source. The clash of continent and ocean floor along the southern margin of remote Alaska threatens people throughout much of the Pacific Rim.

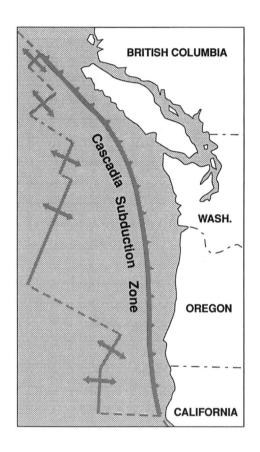

The Cascadia subduction zone has a potential for generating extraordinarily powerful earthquakes along the Pacific Northwest coast. Evidence of rapidly buried soils and marshes along the coastlines of Washington, Oregon, and northern California suggest that several times during the last few thousand years great earthquakes have triggered widespread subsidence of extensive coastal areas. Large sea waves have repeatedly swept inland, inundating many bays, harbors, and estuaries.

Chapter 9

GIANT EARTHQUAKES:
A Possibility for the
Pacific Northwest

National polls consistently rank Pacific Northwest cities such as Seattle, Washington and Portland, Oregon, among the nation's most desirable places to live. Their mild climates and spectacular mountain scenery give these evergreen-shaded cities an abundance of natural blessings.

Killer earthquakes occasionally strike the Puget Sound region, as in 1949 and 1965, but most people do not think of Seattle, Tacoma, or Portland as cities waiting to share the fate of San Francisco. They may be wrong. Recent studies suggest that the Pacific Northwest may be vulnerable to earthquakes of far greater destructiveness than any that have shaken California during historic time. The coastal regions of Oregon, Washington, and British Columbia may experience quakes that will score 9 or higher on a newly devised Moment Magnitude scale for measuring exceptionally large earthquakes. The well-known Richter scale, introduced in 1935 by Charles F. Richter, becomes inaccurate in measuring the strength of seismic events greater than magnitude 8.0.

A Moment Magnitude of 9 means at least ten times the energy release of the 1906 San Francisco earthquake. If an earthquake of that size were centered near the heavily urbanized Puget Sound region it could produce the worst natural disaster in United States history.

The Northwest's potential for enormous earthquakes lies in its geologic setting. From northern California to southwestern British Columbia the eastern edge of the Pacific Ocean floor is sinking under the western edge of North America. The slab of oceanic crust, the Juan de Fuca Plate, is slipping beneath the continental margin at the rate

Damage from the moderate 1965 earthquake centered near Seattle, Washington was widespread but unevenly distributed throughout the Puget Sound region. Local subsidence and sliding literally took the ground from beneath this railroad track in western Washington. —U.S. Geological Survey, McCollough photo

of two or three inches per year. The line of collision, the Cascadia subduction zone, extends approximately 800 miles from Cape Mendocino in northern California to central Vancouver Island, British Columbia.

The Cascadia subduction zone resembles similar regions of ocean floor sinking along the coasts of Alaska and South America. The Alaskan subduction zone has produced some of the largest earthquakes ever recorded, including the 1964 upheaval, which is rated a Moment Magnitude (M) of 9.2. In 1960 subduction off South America triggered the mightiest earthquake of the twentieth century, devastating southwestern Chile and registering a phenomenal M reading of 9.5 (8.5 on the Richter scale). Subduction of the Juan de Fuca Plate beneath the Pacific Northwest has kept the Cascade volcanoes like St. Helens supplied with fresh magma, but it has not produced a high magnitude quake since Europeans first explored the region in the late 1700s.

Earth scientists are not sure why no great quakes comparable to those in Alaska or South America have shaken the Pacific Northwest during the last 200 to 300 years. It may be that the Pacific Northwest subduction zone works smoothly. The Pacific plate may sink steadily instead of in sudden jerks that cause great earthquakes. On the other hand, it seems more likely that the Cascadia subduction zone is now locked and accumulating strain that will eventually break the snag and allow the plate to move in a sudden snap that will generate seismic waves of cataclysmic proportions. Some earth scientists believe that the presently quiet interval may culminate in an earthquake as large as the 1960 Chile cataclysm.

Geologists have recently begun to discover evidence of such monstrous upheavals along the Pacific Northwest coast. Brian F. Atwater, of the U. S. Geological Survey, has found deposits indicating that large sections of the Washington shoreline underwent rapid subsidence at least five times between about 300 and 3100 years ago. The evidence suggests that widespread subsidence of up to six or seven feet along the southwestern Washington coast occurred during several great earthquakes in the Cascadia subduction zone. Widely distributed sheets of sand lying atop intertidal marsh deposits further suggest that large sea waves swept far inland after some of the earthquakes.

Other researchers have discovered dead Western cedar trees whose roots lie in wet soils that were abruptly buried by submergence along the coast of Washington. By counting the annual growth rings of these drowned trees, investigators found that they died simultaneously about 300 years ago, most likely in a large earthquake in which the coastal region subsided several feet. At the same time a huge tsunami, perhaps thirty-five feet high, swept over 125 miles of the Washington and Oregon coast, covering it with a thick layer of sand. Layers of peat lying beneath the sand deposits indicate that the sunken area formerly stood well above the tideline and supported a mature forest. As in the Alaska earthquake of 1964, past Cascadia subduction zone quakes have uplifted some areas while submerging others beneath the sea. Bainbridge Island, three miles west of Seattle, suddenly rose as much as twenty-three feet about 1700 years ago. A nearby area uplifted similarly during one or more major earthquakes about 600 years later.

Gradual, quiet uplift of the Pacific Northwest coast apparently takes place during extended intervals between large earthquakes. One study reveals that a section of the Oregon coast is steadily rising and is now a foot higher in elevation than it was sixty years ago. Such steady uplift creates considerable strain along the subduction zone that will inevitably be released by one or more great earthquakes.

Evidence of great earthquakes that caused large segments of the coast to sink has been found from near Eureka in northern California to Neah Bay in northwest Washington. Similar quakes in the future may sink parts of low-lying towns such as Astoria, Oregon, or Aberdeen and Hoquiam, Washington. Cities farther inland, such as Seattle, Tacoma, and Portland, would experience the kind of intense shaking that damaged nearly every major structure in Anchorage in 1964. The Northwest's most populous cities stand at or near sea level on soft sediments full of water that are likely to liquefy during a severe earthquake. Ground subsidence, surface fracturing, and landsliding could seriously damage highways, powerlines, railroads, and other lifelines as well as commercial and industrial structures. Like San Francisco after a great quake on the San Andreas fault, Seattle could

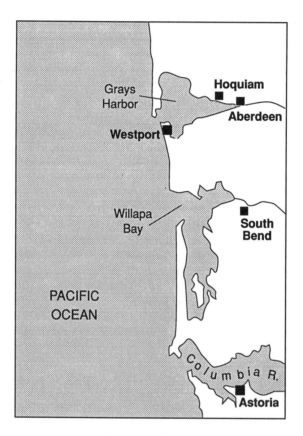

The coasts of southwestern Washington and northwestern Oregon are particularly vulnerable to widespread land subsidence, flooding, and damage from tsunamis triggered by large earthquakes along the Cascadia subduction zone. Low-lying towns like Aberdeen and Hoquiam on Grays Harbor, settlements on Willapa Bay, and Astoria, Oregon, occupy sites that were repeatedly inundated during great earthquakes of the recent geologic past.

find itself temporarily cut off from ground communication with the rest of the world. North-south highways passing through the city, such as Interstate 5, could suffer extensive damage.

Many buildings along Seattle's waterfront and the Pioneer Square area are of unreinforced masonry likely to collapse during severe shaking. Much of Tacoma's business district consists of century-old brick structures that do not meet earthquake-resistant standards. The sea level tidal flats on which Tacoma's industrial and port facilities sit will probably suffer extensive liquefaction and subsidence.

Intense shaking will not only cause widespread sliding and ground displacement along the densely populated shores of Puget Sound, it may also trigger large waves that could sweep through city waterfronts like water sloshing out of an oversize bathtub. Damage to shipping and fuel storage facilities would be severe. Toxic spills, combined with ruptured gas mains and oil storage tanks, could produce a holocaust that both poisons the environment and incinerates crucial structures.

Because Oregon has not experienced a major quake since it was first settled, the state's building codes are less stringent than they are in the Puget Sound area. As a result, Portland has many old masonry structures highly vulnerable to earthquake damage.

In the recent geologic past, huge landslides in the Columbia River Gorge just east of Portland have repeatedly dammed the river. Some of these, such as the Cascade Slide against which Bonneville Dam abuts, may have been triggered by subduction earthquakes. Repetition of large-volume slides could not only damage existing hydroelectric dams and block highways and railroads but also result in widespread flooding downriver after slide-impounded reservoirs break through the slide debris.

Alarmist predictions that the California coast west of the San Andreas fault would disappear beneath the sea are ill-founded, but a similar scenario may apply to parts of the Pacific Northwest. Western California will not drop out of sight, primarily because movement along the San Andreas is horizontal. In the Pacific Northwest the Juan de Fuca Plate is sinking beneath the continental margin. If the subducting plate is now locked and suddenly breaks free, its rapid descent may trigger an extensive subsidence of the Northwest coastline.

Although earth scientists do not expect the towns near Grays Harbor and Willapa Bay on the coast of Washington to become the drowned citadels of a Cascadian Atlantis, they are distinctly vulnerable to a limited coastal subsidence and to the tsunamis that may follow a great earthquake. A recent study of the consequences of an abrupt drop in ground level averaging six feet in Hoquiam shows that virtually all of the city's central commercial and industrial districts would be submerged. Secondary results could include serious air contamination from damage and inundation of a chemical plant. The sudden release of hazardous materials, such as chlorine and ammonia, stored in waterfront areas along the Washington and Oregon coasts could contaminate the air, soil, and groundwater, poisoning many survivors.

During the last few centuries, tsunamis have swept in waves up to thirty-five feet high along the Pacific Northwest coast. In the past their destructive power was mitigated by large stands of trees, which helped to retard the waves' advance inland. Wholesale clearcutting of the coastal forests has largely removed that natural buffer. Future tsunamis will race unimpeded up rivers and estuaries.

Although the Pacific Northwest's greatest earthquake threat stems from the interaction of the North American and Juan de Fuca plates, the region also faces seismic hazards from two other sources. In December 1872 a large earthquake that originated in the continental crust near central Washington, perhaps beneath the Cascade Range, caused severe shaking throughout much of the state. That quake triggered a landslide near Vantage in eastern Washington that temporarily dammed the Columbia River. Other major earthquakes, such as those of 1949 and 1965, come from the subducted eastern edge of the Juan de Fuca plate. Neither of these earthquakes caused as much surface damage as their strength would suggest, primarily because they originated at least forty to fifty miles underground within the deeply embedded plate.

An ancient proverb states that the fortunate man has frequent bad luck on a small scale. The person who lives trouble free may be the gods' target for truly colossal misfortune. The Pacific Northwest may continue to experience moderately damaging shocks like those of 1949 and 1965. Or long centuries of quiet convergence along the Juan de Fuca and North American plates may end abruptly in one catastrophic jolt. Although great earthquakes do not follow a timetable, some earth scientists believe that the present rate of deformation along the Pacific Northwest coast is producing a strain that cannot continue to build for many more decades. Evidence from past disasters makes it increasingly clear that western Washington and Oregon are at extremely high seismic risk. Dating of prehistoric coastal submergences indicates that the Cascadia subduction zone produces one great earthquake about every three hundred years— exactly the period of time that has elapsed since the last one in A.D. 1690.

Light-colored, braided outwash channels mark an alluvial fan in Death Valley. Salt flats in middleground. —Donald Hyndman photo

Chapter 10

THE EARTH STRETCHES ITS SKIN:
Death Valley and the Great Basin

Hiking through Death Valley, California, is like trekking across the surface of Mars. The signs of water erosion are everywhere—stream-cut canyons, broad alluvial fans, deeply incised gorges—but, over vast sun-baked stretches, not a drop of water remains in sight. Like the martian landscape, Death Valley is an intensely arid desert of naked rock haunted by the ghosts of long-vanished floods.

Now bone-dry, the many canyons that indent Death Valley's steep walls show that torrents of water once flowed there. The valley floor itself, encrusted with thick deposits of salt, reveals that a huge lake formerly occupied this parched and seemingly lifeless flat. An earthly counterpart of the red planet's waterless terrain, Death Valley is a geological paradox: the hottest, lowest, and least rained-upon spot in the Western hemisphere, it has a landscape largely shaped by the power of running water.

During the Ice Age when the surrounding mountains bore heavy snowpacks, meltwater repeatedly formed large lakes on the valley floor. The latest to form, Lake Manly, was about ninety miles long, six to eleven miles wide, and had a maximum depth of more than 600 feet. Today's visitors can see the lake's ancient wave-cut beaches, horizontal terraces high on the valley walls at Shoreline Butte and Mormon Point. Lake Manly dried up about 10,000 years ago, but a brief period of cooler, wetter weather permitted a much smaller pond, about thirty feet deep, to accumulate about 2,000 years ago. When that pond evaporated, it left behind a surface layer of salt which wind and rain subsequently sculptured into the Devil's Golf Course.

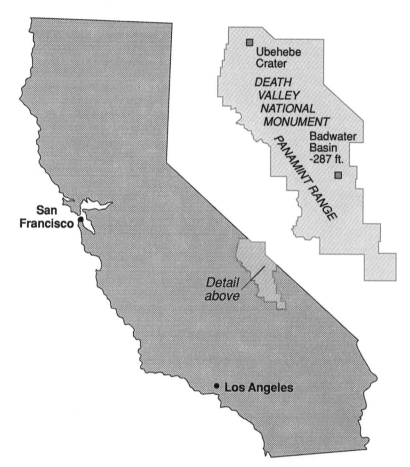

Map showing location of Death Valley National Monument, the lowest point in the Basin and Range Province. Ubehebe Crater is one of the younger cinder cones in the Basin and Range Province.

Although on a much smaller scale, water continues to reshape Death Valley's treeless contours. The annual precipitation averages only 1.71 inches at Furnace Creek, but it commonly falls in short, intense downpours. The heat-baked surfaces, that lack vegetation, do not absorb the rain, so runoff triggers flash floods that rush through normally dry channels to deposit more silt and gravel on the valley floor. A single winter storm can quickly transform Death Valley from a waterless desert to a floodplain in which its few roads become temporarily impassable beneath raging streams and flowing mud.

During the last few million years, such seasonal floods, great and small, along with other erosional processes, have gradually blanketed the valley floor with sediments nearly two miles thick. Such a phenomenal accumulation would have filled an ordinary valley to the brim, but Death Valley is not a ordinary stream-cut valley. It is a dropped block of the earth's crust that is still sinking faster than material eroded from the surrounding mountains can fill it. The structure of Death Valley is typical of the geological province to which it belongs. At 282 feet below sea level, it is the deepest part of the Great Basin, a region, mostly in Nevada, from which streams find no exit but drain internally into closed basins. The Great Basin lies within the Basin and Range Province. This vast area of rugged mountains alternating with deep troughs stretches from Utah's Wasatch Range on the east to the Sierra Nevada on the west. From north to south, the Basin and Range extends from southeastern Oregon and southern Idaho through southern New Mexico and Arizona to north-central Mexico.

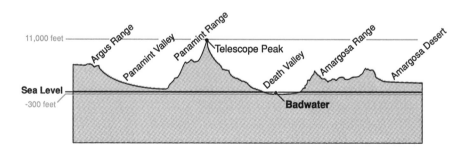

This cross section of the Death Valley region illustrates the dramatic differences in elevation between neighboring locations in the Great Basin. Telescope Peak in the Panamint Range stands more than 11,000 feet above the Badwater depression, at 282 feet below sea level the lowest point in North America

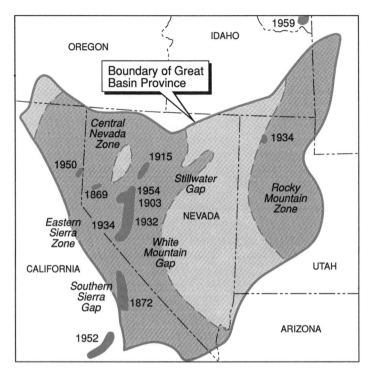

Earthquake hazard zones in the Basin and Range Province. Note that most large historic quakes accompanied by surface faulting happened along a belt about 125 miles wide that extends from California through central Nevada. The largest shock, approximately magnitude 8.0, occurred in 1872 on the Owens Valley fault near the east front of the Sierra Range. The 1952 Kern County earthquake, the largest in California since 1906, took place along the same trend but outside the province to the southwest. The 7.1 Hebgen Lake earthquake occurred in 1959 at the northwest extremity of the seismic zone.

The Basin and Range is a region of north-south trending, parallel mountain ranges separated by long enclosed depressions. Earth scientists believe that this topography results from uplift and stretching of the continental crust that began about 17 million years ago. As the crustal rocks were stretched thin, faults split the crust into numerous blocks. Movement along faultlines raised some blocks to form mountain ranges, while others sank to form depressions.

Movement along the Basin and Range faults continues. As the accompanying map indicates, most of the region's historic earthquakes are concentrated along a north-trending belt approximately 125 miles wide that extends from California through central Nevada. The largest quakes, registering magnitudes of 7.0 to 8.0, occurred in

1872, 1915, 1932, and 1954. The most severe California quake since 1906, the Kern County or Tehachapi earthquake of July 1952 took place outside the Great Basin but along the same trend.

On average, the region's large earthquakes take place about every twenty-two years. The most recent was the magnitude 7.1 Dixie Valley, Nevada, quake of 1954. Because more than thirty-five years have elapsed since that event, some earth scientists believe that the Great Basin is overdue for another sizable shake. Future earthquakes are expected to occur along faultlines that are geologically young but have not moved during historic time, such as the Stillwater, White Mountain, and Southern Sierra gaps.

The Wasatch fault zone, near which eighty-five percent of Utah's population live, has not produced a notable earthquake since the area was settled in 1847. Studies of the fault's history suggest that the average recurrence interval for moderate-to-large earthquakes in the Wasatch fault zone is between fifty and 400 years. Salt Lake City thus joins San Francisco, Los Angeles, Seattle, and other western cities in preparing to cope with a major earthquake.

The most powerful quake recorded in the Basin and Range region originated at its western extremity where the extending crust of the continental interior encounters the immense granite solidity of the Sierra Nevada. The seemingly near-vertical eastern face of the Sierra represents North America's largest fault scarp, graphic evidence of earthquakes that have elevated the southern range to its present height. Faults associated with the Sierra uplift and dropping of the adjacent Owens Valley generated the great earthquake of 1872. That quake produced a vertical displacement of twenty-three feet between the Sierra Nevada and Owens Valley, the westernmost basin in the province.

Seismic activity is also intense along the eastern margin of the Basin and Range Province, which continues to expand eastward as the North American plate drifts to the west over deep-seated rifts in the crust. Single large earthquakes cause valley floors to drop or mountains to rise as much as ten or twenty feet at one time. On August 17, 1959, a magnitude 7.1 quake, centered near Hebgen Lake, Montana, northwest of Yellowstone National Park, was felt over 600,000 square miles. The Hebgen Lake shock triggered an avalanche that took twenty-eight lives and dammed the Madison River, forming the appropriately named Earthquake Lake. Idaho's worst earthquake struck the eastern part of the state on October 25, 1983, killing two children in Challis. Felt over 330,000 square miles, the magnitude 7.0 shock destroyed property worth $12.5 million. The earthquakes of

1925 and 1935, which caused severe damage in Helena, Montana, also represent fault movement along the eastern edge of the expanding Basin and Range.

Although Death Valley has experienced only minor earthquakes during historic time, geologists expect the faultlines that border it to generate larger quakes in the future as bedrock beneath the valley floor continues to sink. As crustal extension continues to enlarge the Basin and Range area, earthquakes will occur sporadically throughout the region, but especially at its eastern and western margins.

Volcanoes, as well as earthquakes, have helped to shape the Basin and Range topography. Lava flows and small hills of fragmental rock, cinder cones, dot the landscapes of Nevada, southern Idaho, Arizona, and eastern California. Some of the region's latest volcanic activity took place at Ubehebe Crater near the northern end of Death Valley. Ubehebe Crater, Little Hebe, and the neighboring cluster of small cones grew only a few thousand years ago during eruptions that buried the surrounding terrain under a blanket of basaltic ash as much as 150 feet thick.

Aerial view of Panamint Valley, a typical Basin and Range terrain. Like its near neighbor, Death Valley, Panamint was created by the downdropping of a crustal block between two parallel uplifted blocks, the Argus Range on the right and the Panamint Range on the left. Deep crustal fractures, fault zones, mark the boundaries between the blocks. —California Bureau of Mines and Geology photo

View across salt flats to the Panamint Mountains, southern Death Valley, California. —Donald Hyndman photo

Ubehebe Crater, which is about a half mile in diameter and nearly 800 feet deep, was created largely by phreatic or steam blast explosions. This type of eruption occurs when subterranean magma encounters groundwater, which flashes into steam and blasts open a crater surrounded by a relatively thin rim of volcanic ejecta, a maar. A short hike to the lip of Ubehebe Crater reveals that much of its inner wall is composed of the alluvial sediments that veneer this part of the valley.

While Ubehebe and its fellow maars are young enough to retain their original forms, older and far larger volcanic deposits in Death Valley have been extensively modified by erosion and fault movement. The visitor to Artist Drive may marvel at the brilliant colors displayed in the huge blocks of welded rhyolitic ash that have been downfaulted from the adjacent Black Mountains. Capped by dark basaltic lava flows, these enormous rhyolitic blocks reflect the desert sun in hues ranging from bright orange and rust to the subtlest pinks, grays, and other pastels. Like the volcanic rocks in Yellowstone, the Artist Drive deposits owe their colors to the effects of hot water percolating through volcanic rocks.

Although little known to most Americans, the Basin and Range region is geologically young and active. With its high potential for large earthquakes and volcanic eruptions, it is increasingly likely to make its presence felt.

85

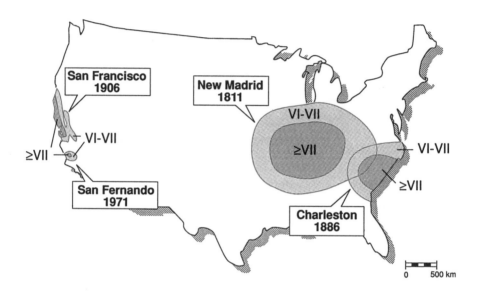

Comparison of areas of the United States shaken with damaging intensity in four large earthquakes. Note that the effects of ground shaking extend over a much larger area in the eastern states than in the West. The distributions of the intensities (Modified Mercalli VI or greater) of the 1811 New Madrid, Missouri, earthquake (about 8.0 on the Richter scale) and the 1886 Charleston, South Carolina earthquake (about 6.6) each were substantially greater than those of the 1906 San Francisco earthquake (Richter 8.3) and the 1971 San Fernando, California earthquake (Richter 6.4). The large rigid crustal plate underlying the eastern U.S. transmits seismic waves more efficiently than the deeply fractured and faulted crustal blocks composing the western region. (After Schnell and Herd, 1983)

Chapter 11

EARTHQUAKES IN
UNEXPECTED PLACES:
The Catastrophic
New Madrid Quakes of 1811-1812

In late 1983 Bhagwan Shree Rajneesh—then the Rolls-Royce collecting leader of a religious commune in central Oregon—warned his California followers to leave the state by the year's end. The Bhagwan predicted that 1984 would introduce a fifteen-year-long series of natural and man-made catastrophes, including earthquakes, volcanic eruptions, violent storms, and a nuclear holocaust. Among the population centers he scheduled for decimation were San Francisco, Los Angeles, New York, Bombay, and Tokyo.

The Bhagwan's year of disaster, 1984, is long gone, but almost every succeeding new year brings similar forecasts of doom from various clairvoyants. Most such prophets focus on California, where mystics and channelers regularly announce the Big One's imminent arrival.

The absolute certainty of future large earthquakes in California is common knowledge. We need no swami or soothsayer to warn us of a hazard that earth scientists have long labored to confirm and define. What the Bhagwan and his ilk apparently overlook is that states other than California are the more likely sites of truly catastrophic earthquakes. The largest series of earth shocks in American history was centered near New Madrid, Missouri. Between December 16, 1811, and the spring of 1812 a swarm of appallingly destructive earthquakes convulsed the central Mississippi Valley, triggering massive landslides, submerging whole islands, opening vast fissures in the earth, and stirring the great river's surface to a boiling fury.

Craterlet formed by sand blow during the August 31, 1886 earthquake centered near Charleston, South Carolina.—U.S. Geological Survey photo

Three factors distinguish the New Madrid earthquakes from other historic shocks. First, they caused damage over an exceptionally large area and were felt over even greater distances. Destruction was severe throughout southeastern Missouri, northeastern Arkansas, southern Illinois, and western Kentucky, Indiana, and Tennessee. Because settlements in the affected area were few and small, casualties and economic losses were modest.

No seismographs existed then, so the earthquakes' magnitudes are perhaps best measured by their awesome effects on the land itself. Large areas rose and even larger sections sank, transforming dry land into lakes or swamps. Tennessee's Reelfoot Lake, eighteen miles long, two to five miles wide, and five to twenty feet deep, was created or greatly enlarged during the earthquakes. Lowlands flooded and the earth's surface was split into numerous yawning fissures many feet across. Plumes of sand spurted out of the ground like geysers, rising tens of feet into the air and producing strong sulphurous odors from decayed vegetable matter previously buried underground.

The vast distances over which the earthquakes were felt are equally remarkable. Tremors were reported as far north as Canada, in Detroit, 600 miles from the epicenter, and in Boston, 1,100 miles to the northeast. As if anticipating its invasion and burning by the

British during the impending War of 1812, Washington , D. C., 700 miles east of New Madrid shuddered ominously, as did New Orleans, 500 miles to the south. Approximately one million square miles, half the area of the entire United States, trembled convulsively.

The large number of separate, powerful shocks is also unparalleled in this country's seismic history. In most historic earthquake sequences, the initial jolt is followed by a few lesser tremors of generally diminishing intensity. The New Madrid earthquakes, however, continued at a high level of intensity for many months, with the greatest convulsion occurring on February 7, 1812, nearly eight weeks after the first violent shock. One particularly acute observer, Jared Brooks, an engineer and surveyor of Louisville, Kentucky, devised an instrument to measure the relative size and severity of the quakes. Using pendulums to record horizontal movement and several springs to detect vertical movement, Brooks registered 1,874 distinct shocks between December 16, 1811, and March 1812. He and other observers agreed that the shakes of December 16, January 23, and February 7 were the worst, although Brooks also classified five others during this period as "tremendous," ten more as "severe," and an additional thirty-five as "alarming to people generally."

The third peculiarity of the New Madrid earthquake series is the unusual length of time during which the quakes persisted. Around New Madrid, the earth shook at least once every day through the end of 1812. With time the disturbances apparently migrated northward and continued for an additional two years. Between 1812 and 1814 earthquakes were an almost daily event at Saline Mines, Illinois. A recent study comparing the number and magnitude of the New Madrid earthquakes between December 16, 1811, and March 15, 1812, with those that occurred in southern California between 1932 and 1972 shows that in three months the New Madrid shocks equalled those in California during a forty-one year period.

The precise cause of the New Madrid earthquakes remains conjectural. Unlike the California quakes, most of which occur along recognized faults that rupture the earth's surface, the Mississippi Valley earthquakes struck an area with no visible surface faulting. Some geologists speculate that the earthquakes originated in a deeply buried rift valley, a block of the earth's crust that dropped about five hundred million years ago but which may remain sporadically active.

As a 1982 United States Geological Survey study reveals, between 1838 and 1976 at least twenty damaging earthquakes occurred in the central Mississippi Valley, apparently centered at the northern or southern end of the New Madrid fault zone. In 1895 an estimated magnitude 6.2 shock, its epicenter near the junction of the Mississippi and Ohio rivers, was felt over one million square miles in twenty-three states and Canada. Besides causing considerable damage near Charleston, Missouri, it created a new four-acre lake. A 1909 earthquake, centered near Aurora, Illinois, broke chimneys in the Chicago suburbs forty miles away. Although relatively minor, it also was felt over an astonishingly wide area—half a million square miles. Earthquakes of central and eastern North America shake such extremely large regions because the underlying rock is old, cold and brittle. Relatively few faults break the ancient bedrock so that seismic waves travel great distances with little diminution of intensity.

California may well endure another devastating earthquake before the central Mississippi Valley is again visited by shocks comparable to those of 1811-1812. But purveyors of seismic calamity like Bhagwan Shree Rajneesh and his counterparts must remember that our ostensibly stable midcontinent has a significantly greater potential for human tragedy than the media-glamorized Golden State. Today approximately thirteen million people live in the area most severely shaken by the New Madrid earthquakes, compared to about six million persons who inhabit the northern California terrain afflicted in 1906. The present combination of high population density and prevalence of buildings that do not meet earthquake safety standards could result in widespread damage to property and an extremely high death toll. A recent survey reveals that St. Louis County alone has 140,000 unreinforced masonry structures. Four out of five city buildings are also unreinforced masonry, the kind most likely to collapse during an earthquake. According to one estimate, a replay of the New Madrid earthquakes would cause property losses exceeding $50 billion.

St. Louis, Missouri and Cincinnati, Ohio experienced only cracked brick walls and other slight damage in 1811. The effects of future earthquakes could be far more serious. Memphis, Tennessee, which did not exist at the time of the New Madrid quakes, occupies a site that churned with earth-splitting intensity. If the townsite were struck a comparable seismic blow today, frame buildings would be thrown off their foundations, while the city's tens of thousands of brick buildings would crumble into piles of rubble. Many towns and cities in the New Madrid seismic hazard zone, including parts of Memphis and St. Louis, sit atop wet, sandy soils of the Mississippi River floodplain. The epicenters of future large earthquakes is likely to be as close to these areas as the 1989 Loma Prieta quake was to San Francisco, where ground shaking was greatly intensified as seismic waves rolled through water-logged landfill.

90

Many buildings in Charleston, South Carolina were badly damaged or collapsed during the August 31, 1886 earthquake. The nature of the damage varied widely, ranging from toppled chimneys and fallen plaster to total structural collapse. —U.S. Geological Survey photo

Earthquake hazards can not be evaded simply by moving from Missouri or Tennessee to the East Coast. Much of the eastern United States also lives under an earthquake threat. In August, 1886, a violent trembler struck Charleston, South Carolina, destroying or damaging most buildings in the area and killing sixty people. Although its Richter magnitude is estimated at only 6.6, the Charleston earthquake set the ground in motion throughout most of the eastern United States. Minor damage to upper floors of buildings occurred as far away as New York City, 600 miles, and Chicago, 720 miles.

The 5.2 magnitude trembler that rattled through upper New York state on October 7, 1988, reminded several million persons that the northeastern seaboard is also seismically at risk. Felt from Montreal, Canada, to New Jersey, it was the largest quake felt in the area since a 5.9 event in 1944. Since early colonial times, New England and other northeastern states have been visited by random shocks. The first major shake recorded took place in 1683; another struck Massachusetts in 1755, when chimneys toppled and church bells clanged wildly in Boston. New York state alone has experienced more than 350 sizable quakes since 1720. Although occurring less frequently than in California or Alaska, such tremors remind us that that a truly destructive jolt could strike the most densely populated region in the United States.

Long a Charleston landmark, St. Philip's Church sustained damage to its picturesque bell tower during the August 31, 1886 earthquake. Note the partly collapsed tile roof in the building on the right and fallen bricks littering the street. —U.S. Geological Survey, C. C. Jones photo

The results of a major earthquake centered near a large metropolitan area in the eastern United States may be even more catastrophic than a magnitude 8-plus in California. In crowded eastern cities, millions of people live in old brick and unreinforced masonry buildings that are likely to collapse during the intense shaking that will affect perhaps a million square miles. In the almost complete absence of adequate state or local planning for such an event, services to provide medical help for the injured, feed and shelter millions of homeless survivors, and restore broken water, electric, gas, transportation, and communication lines will be almost entirely lacking. Society's inability to cope with the disaster may compound the suffering and have social consequences that will last for decades.

While millions of Californians consciously dread the coming quake, most people living in the eastern United States remain blissfully unaware that Memphis, New York, or Charleston may soon experience the kind of geologic violence that ravaged San Francisco in 1906 and Anchorage in 1964. Thus far in our history we have been fortunate that most large quakes took place either in remote areas or, as was the case with the New Madrid shocks of 1811-1812, when the United States had a much smaller population. The timing of large earthquakes has also been remarkably fortuitous: the California shocks of 1857, 1872, 1906, and 1971 occurred either at night or during the early morning hours when most people were safely in bed. Both the Long Beach earthquake of 1933 and the Alaska upheaval of 1964, which demolished many public school buildings, happened in the late afternoon after schools had emptied and most commuters had left downtown areas. Even the Loma Prieta earthquake, which occurred at 5:04 p.m. during a normal urban rush hour, took place on a day when many people had gone home early to watch the televised 1989 World Series, ironically pitting San Francisco against Oakland. We can not expect nature to time future earthquakes with such consideration for human welfare.

Increased public awareness that Missouri, Tennessee, South Carolina, Massachusetts, and New York—as well as California and Alaska—are vulnerable to powerful earthquakes is a life-saving necessity. An informed citizenry may persuade government leaders to revise building codes, plan for emergency services, and otherwise prepare to deal with the harsh social consequences of geologic reality.

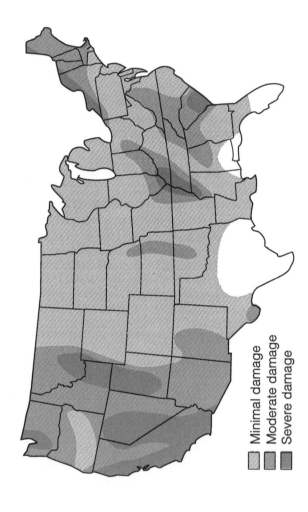

Except for a few regions along the Atlantic Gulf, virtually every part of the United States is subject to earthquake risks. Based on the historical record, areas where the worst shaking is likely to occur (red zones) include the Puget Sound region in Washington state, most of California except for the Central Valley, western Nevada, the western Rocky Mountains, the central Mississippi Valley, the southeast seaboard, and the New England states. This map vividly pinpoints regions in which uniform codes for earthquake-resistant buildings can significantly reduce potential losses of life and property.

—Modified from Algermissen and Perkins, Jaffe et al., 1981

Minimal damage
Moderate damage
Severe damage

Chapter 12

WHERE AND WHY DESTRUCTIVE EARTHQUAKES WILL STRIKE NEXT

Terrified by the violence of Alaska's 1964 earthquake, a young wife made her husband promise to resettle in an area where earthquakes almost never occur. The couple chose Sacramento, a tree-shaded city in the middle of California's broad Central Valley. Despite its low-key midwestern ambience, California's capital is not immune to earthquakes. Although it was only mildly shaken by the San Francisco quake of 1906, it was rudely jolted in 1892 by movement along a fault on the western edge of the valley. Other faults, as yet unknown because they lie buried deep beneath sediments thousands of feet thick, may riddle the valley bedrock. Some may produce earthquakes that could upset Sacramento's usual placidity.

Except for parts of Texas, Florida, and some other Atlantic Gulf states, Americans fearing earthquakes can find few seismic-free refuges in the United States, including such vacation meccas as Hawaii, Puerto Rico, and the Virgin Islands. Although most Hawaiian quakes are minor blips associated with volcanic activity, some are powerful enough to trigger giant sea waves that devastate coastlines. After the earthquake of April 2, 1868, a large tsunami reportedly swept over the tops of trees along the Big Island's south shore. Vacationers should remember that such destructive waves will accompany most future earthquakes in Hawaii.

The Caribbean resorts of Puerto Rico and the Virgin Islands also offer no escape from damaging quakes. These tropical retreats are part of a seismic zone that intermittently experiences strong temblors. In 1918 a magnitude 7.5 shock killed 116 persons in Puerto Rico

The San Fernando, California earthquake of February 9, 1971 scored only 6.6 on the Richter scale, but caused hundreds of millions of dollars in damage. The collapse of the overpass at Interstate 5 and State Highway 14 vividly demonstrates the destructive power of a "moderate" shock. Since 1971 California has attempted to make its freeway bridges quake-resistant, an effort that was dramatically tested during the 1989 Loma Prieta earthquake. —U.S. Geological Survey photo

and was accompanied by a large tsunami. Although great earthquakes have not struck the region during historic time, some geologists note that subduction of the Caribbean plate is analogous to that along the western end of the Aleutian Islands and may eventually produce a catastrophic shaking and huge sea waves that may devastate some of the world's most popular resorts and beaches.

Our continental interior is no less at risk. Several Rocky Mountain states, as well as the vast Basin and Range region, particularly its expanding eastern and western extremities, are subject to violent shaking. The Wasatch Mountain front at Salt Lake City may soon produce a devastating shock.

Most of the world's mightiest earthquakes occur along subduction zones like those bordering western South America or Alaska, where tectonic plates carrying continent and ocean floor collide. The ominously quiet Cascadia subduction zone may soon trigger a gigantic quake that will rock the entire Pacific Northwest. While large segments of the coastline sink six or seven feet and tsunamis surge inland, Puget Sound may send massive walls of water crashing into the Seattle waterfront. Thousands of old masonry buildings in Bellingham, Seattle, Tacoma, Portland, and Eugene may explode in showers of bricks, burying cars and pedestrians and rendering downtown streets impassable. Desperately needed relief services may not reach inundated coastal areas for days because major highways will be severely damaged by landslides and widespread ground failure.

Californians who live on or near the world's most notorious fault, the San Andreas, also have reason to worry. In 1989 the Loma Prieta earthquake staged a small preview of the widespread destruction that scientists expect when the northern San Andreas generates another magnitude 8.3 shock, an event that has a fifty-fifty chance of happening within the next thirty years. The odds are even higher that the south-central segment of the fault, which has not produced a great quake since 1857, will soon convulse much of southern California, perhaps before the end of the century.

An electric transformer thrown down by the 1971 San Fernando quake. Anticipated major quakes on the San Andreas, Hayward, and other California fault zones are expected to rupture vital communication and lifelines, leaving millions of persons without electrical power for at least several days. —U.S. Geological Survey photo

Lesser known faults that slice through densely populated areas threaten equal or worse disasters. The Newport-Inglewood fault that produced the lethal 1933 Long Beach earthquake may cause even more deaths and higher property losses in the Los Angeles metropolitan area than the San Andreas, which is located farther from southern California's population centers. A 7.5 magnitude trembler on the Hayward fault may create greater damage in the San Francisco Bay area than an 8.3 repeat of the 1906 quake on the northern San Andreas.

As California's population grows, moderate quakes will take an increasingly heavy toll. The 1971 San Fernando earthquake had a modest 6.6 magnitude, but it killed 65 people, injured many others, and caused $1 billion in property damage in the Los Angeles area. The 1983 Coalinga quake, magnitude 6.7, virtually leveled the downtown area, injured 45 people, and caused $31 million in property damage. The Loma Prieta temblor, over thirty times weaker than the anticipated Big One, ranks as the single most costly natural disaster ever to strike our nation.

The Parkfield Forecast

Fears of California's Big One naturally focus on heavily settled areas, but seismologists are more confident about forecasting the size and timing of a moderate earthquake on a sparsely inhabited section of the San Andreas fault. In 1985 the U. S. Geological Survey predicted that an earthquake of approximately magnitude 6.0 would occur before 1993 on the San Andreas Fault near Parkfield, a tiny hamlet (population 34) in west-central California. The first officially recognized scientific prediction of an earthquake in the United States, the USGS forecast is based largely on the remarkable regularity with which moderate temblors have struck Parkfield during the last century.

Located on the San Andreas about mid-way between San Francisco and Los Angeles, Parkfield experiences a magnitude 6.0 trembler about every twenty-two years. After the great Fort Tejon event of 1857, Parkfield was shaken in 1881, 1901, 1922, 1934, and 1966. On the last two dates, foreshocks of magnitude 5.0 occurred exactly seventeen minutes before the main quake. Instrumental monitoring of this segment of the San Andreas makes it the best-studied fault zone in the world. If the next quake's timing and magnitude materialize as predicted, it may lead to enhanced prediction capability on other segments of the fault.

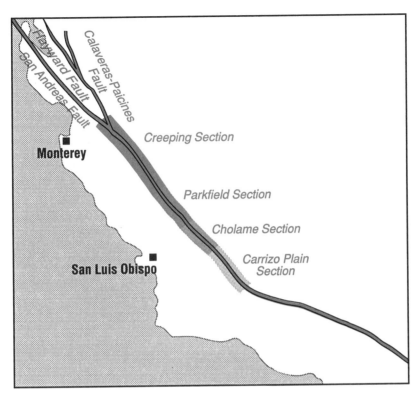

The U.S. Geological Survey believes that a 6.0 magnitude earthquake will occur on the Parkfield section of the San Andreas fault before 1993. Since about 1900, quakes of that size have shaken the Parkfield area about every twenty-two years, the last in 1966. The Cholame and Carrizo Plain sections of the San Andreas ruptured during the great earthquake of 1857 and may trigger another catastrophic earthquake in Southern California within the next thrity years.

Earth scientists studying that part of the northern San Andreas fault that caused the 1989 Loma Prieta earthquake had predicted two years before the event that it would produce a shock as strong as magnitude 6.5 within the next thirty years. The announced time frame was too large to prepare for a specific date, but such general predictions help increase public awareness of seismic threats and encourage adoption of building codes that make structures in hazard zones quake-resistant.

The ability to pinpoint the arrival of an impending earthquake seems desirable, but if earth scientists eventually are able accurately to forecast a quake hours or days in advance, their success may cause more social problems than it solves. Pity the unfortunate governor who must decide what to do with such a forecast. If he announces that a magnitude 8.3 quake is expected to strike Los Angeles before noon tomorrow, will his warning save lives or produce a panic in which many are killed or injured trying to evacuate the city? If an evacuation is ordered and the quake does not occur until after people have returned to their homes and workplaces, who is responsible? Or, if the quake does not take place at all, who must answer for the social and economic disruption?

During ordinary rush hours traffic typically slows to a frustrating crawl on southern California freeways and the few long bridges connecting San Francisco Peninsula with the mainland. Minor auto accidents commonly impede or stop highway traffic for hours—and that on a normal day. After an emergency warning, frightened residents fleeing urban centers where the chance of being crushed by falling buildings is greatest, may suffer almost as many accidents and injuries as they are trying to escape.

While suburban residents can mitigate risks by sleeping in their back yards until the danger is over, many urbanites in highrise apartments or tenements have nowhere to go. Most cities do not have enough parks or other open spaces, or water or sanitation facilities, to accommodate anxious crowds awaiting disaster.

California's San Andreas fault, marked by a white line, runs through Daly City, a suburban community immediately south of San Francisco. Residences built atop or near the faultline can expect shaking of maximum violence during the next major earthquake. Houses perched above steep cliffs overlooking the Pacific Ocean, right foreground, can expect to be engulfed in landslides, an estimated 2,000 of which occurred along the coastline between San Francisco and Santa Cruz during the 1906 shock.
—U.S. Geological Survey, R. E. Wallace photo

View north toward San Francisco along the San Andreas faultline. Since this photo was taken, development in the area has intensified. San Andreas Lake in the top center gives the fault its name. —U.S. Geological Survey, R.E. Wallace photo

Because earthquakes are a permanent feature of life in California, some of its cities lead the nation in taking measures to reduce earthquake hazards by enforcing stringent building codes and preparing state agencies to cope when the inevitable occurs. Despite delays in bringing its older masonry buildings up to code, San Francisco pursues a generally high standard of earthquake preparedness. By contrast, nearby Daly City, has permitted construction of numerous tract developments near or even directly on the San Andreas fault. The East San Francisco Bay cities of Oakland, Berkeley, and Hayward have built hospitals, police stations, and other essential public service structures atop the active Hayward fault. Other parts of the country, where earthquakes are less frequent, typically have ignored seismic risks altogether.

The Eastern United States at Risk

Earthquakes in the eastern United States afflict larger areas than comparable events in the West because east of the Rockies seismic waves travel farther through solid crustal rock and with less diminution of energy. This lower attenuation of seismic waves in the East allows severe shaking to extend over vast distances from the quake's epicenter.

Because most people east of the Rockies have not felt a quake during their lifetime, they exert little pressure on public officials to improve building standards or prepare for the consequences of a major earthquake. As a result, building design and construction do not meet

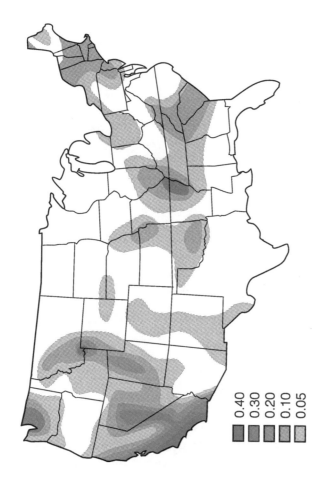

0.40
0.30
0.20
0.10
0.05

This contour map designates areas of the United States in which the highest degree of ground acceleration, violent shaking, is anticipated. The contour lines show ground acceleration expressed as the force of gravity in decimal fractions. The higher the acceleration rate, the greater the anticipated damage to ordinary, non-earthquake-resistant structures. The hazards are particularly acute east of the Rockies, where most buildings are not designed to survive even moderate tremors. —Modified from Jaffe et al., 1981

the earthquake-resistant standards enforced in some western states. When a large shock hits, hundreds of thousands of poorly constructed masonry buildings will collapse, many fatally injuring their inhabitants.

Some earth scientists believe that a major earthquake will strike somewhere in the eastern two thirds of the nation in the next twenty years. It is impossible to pinpoint the exact location, but the central Mississippi Valley states, South Carolina, New England and New York are among the top hazard zones. Robert I. Ketter, director of the National Center for Earthquake Engineering Research at the State University of New York at Buffalo, recently observed that a large eastern city, such as Memphis, Charleston, Boston, or New York, will probably experience a destructive earthquake by the year 2000. While the probability of a high magnitude quake hitting any particular spot is low, Dr. Ketter noted that the probability of one occurring somewhere in the eastern United States is better than 75 to 95 percent. He warned that by the year 2010 the probability increases to almost 100 percent.

Americans must learn to view their earth in geologic terms, recognizing that both our ocean floors and continents are in virtually constant movement, creating earthquakes in both expected and unexpected places. These shocks to human complacency, whether they occur in California, Missouri, South Carolina, or New York, remind us that our planet, with its incandescent interior and thin, ever-shifting crust, pulsates like a living organism. As mere passengers on an evolving globe, we ignore its irregular movements at our peril.

On May 18, 1980, Mount St. Helens devastated more than 230 square miles of prime timberland in southwestern Washington and killed nearly sixty people. At times its cloud of hot ash rose as high as 90,000 feet into the stratosphere, where winds soon carried it around the world. —U.S. Geological Survey photo

Part II:
Volcanic Hazards in the West

Chapter 13

MOUNT ST. HELENS:
A Lethal Beauty

Homeowners in the peaceful, heavily wooded Toutle River valley of southwestern Washington refused to believe it. National Guardsmen in helicopters hovering overhead shouted through megaphones, urging everyone to evacuate the area immediately. St. Helens was erupting and a huge wave of mud, boulders, downed trees, and debris from smashed buildings was moving rapidly downstream toward their homes.

Looking about them at the reassuring familiarity of Douglas fir, tranquil stream, and high protective valley walls, some residents ignored the warning and continued their usual Sunday routines that clear spring day, May 18, 1980. Blocked from their view, a mushroom-shaped cloud, twenty miles across at its base, broiled 80,000 feet into the stratosphere. Caught in a ground-hugging blast of gas and hot rock that had swept northward from the volcano that morning, nearly fifty persons were already dead and more were about to die. Among those at risk were some Toutle River residents who apparently could not conceive of a sudden and cataclysmic change in the natural environment they had known for a lifetime.

A few hours earlier, at daybreak May 18, campers, hikers, and scientists monitoring the volcano had viewed a serene mountainscape. Dawn found St. Helens' once glistening icefields begrimed with gray ash from the minor eruptions that had begun on March 27, while the evergreen forest surrounding Spirit Lake at the volcano's northern foot drooped slightly under its burden of ash. The oversteepened north flank bulged ominously, but the crater was as still as the sunny morning air. It looked as if observers would experience another weekend of uneventful volcano watching.

At 8:32 a.m. chaos suddenly disrupted the familiar natural order. An earthquake, registering 5.1 on the Richter scale, struck directly beneath St. Helens' oversteepened north slope. The volcano's north side, deeply weakened by the body of magma that had slowly risen into the cone during the previous six weeks, broke loose in an enormous landslide. Three quarters of a cubic mile of shattered rock, the largest landslide in recorded history, avalanched north at about 200 miles per hour. Crashing into Spirit Lake, six miles north of the summit, part of the slide overtopped a ridge 1,200 feet high. The main arm of the avalanche traveled approximately seventeen miles west down the North Fork Toutle River, which drains the west end of Spirit Lake.

As the north face of St. Helens peeled away in the landslide, it uncapped the superheated water dissolved in the molten rock inside the cone. Released from confinement, the water flashed into steam, expanding its volume a thousandfold and creating a pyroclastic surge—a deadly mixture of incandescent gas, shredded magma, and fragments of old rock torn from the mountain—that quickly overtook the initial landslide. Traveling faster than the speed of sound and expanding in a fan-shaped arc to the northeast, north, and northwest,

Map showing major features of the paroxysmal May 18, 1980 eruption of Mt. St. Helens. Gray stippled area indicates effects of the pyroclastic surge (blow down area). Mudflows extended down nearly every valley heading on the volcano. —After Lipman and Mullineaux, 1981

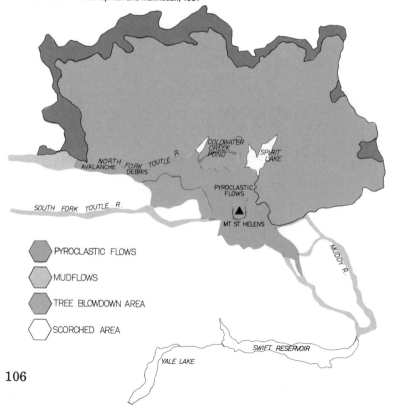

the pyroclastic surge mowed down thick stands of Douglas fir, two hundred feet high, as far as eighteen miles from the summit. The pyroclastic surge was immediately followed by an immense wave of hot ash, which sandblasted forest and topsoil for several miles north of the crater, scouring the ground down to bedrock.

The initial landslide carried large blocks of molten rock, some of which plunged into Spirit Lake and the head of the North Fork Toutle River. When the hot magma hit lake or river water, the explosion was heard as far away as Vancouver, British Columbia. The secondary blasts helped energize the pyroclastic surge with its lethal power to surmount a series of high ridges and decimate 230 square miles of largely pristine lakes and forest.

The eruption continued all day, pumping millions of tons of volcanic ash, pulverized rock, into the stratosphere. High-level winds carried the ashcloud to the east, turning day into night across eastern Washington, northern Idaho, and western Montana. Thousands of square miles were blanketed with pale gray powder the color and consistency of dry cement. Some St. Helens ash would drift around the globe several times before eventually settling to earth.

Because the landslide had removed the entire north side of the cone, the fountains of incandescent magma that shot from the crater poured north into Spirit Lake and the North Fork Toutle River. The seething gas and frothy volcanic glass set river and lake water boiling. Groundwater and blocks of glacial ice trapped below or within the pyroclastic flows flashed into steam that blasted secondary craters along the north shore of Spirit Lake.

Mobilized by melting icefields and river water, voluminous mudflows swept down almost every stream valley heading on St. Helens. The largest traveled down the North Fork Toutle River, joined the main Toutle River valley, then emptied into the Cowlitz River near the town of Castle Rock. Turgid floods continued down the Cowlitz into the Columbia River, where volcanic debris temporarily closed the channel to ocean-going vessels.

Five smaller but still violently explosive eruptions occurred during the summer and fall of 1980, several of which spread ash over Portland, Vancouver, and other cities of northwestern Oregon and southwestern Washington. Except for a few explosive bursts in 1982, 1984, and 1990, most of St. Helens' subsequent activity was confined to extrusions of lava inside the crater. Most of the gas dissolved in St. Helens' current magma supply had escaped during the 1980 activity, so the following series of eruptions were relatively quiet affairs in which degassed magma oozed out in thick tongues of lava. Stiff and pasty, this dacite magma piled up to form a huge lava dome, over 3,500 feet in diameter and 925 feet high.

Earth scientists from the U. S. Geological Survey's newly established Cascades Volcano Observatory in Vancouver, Washington, were able to predict in advance every eruptive episode after mid-1980 and to issue public warnings. When magma began to move up through the volcano's plumbing, seismographs recorded the earth tremors and tiltmeters registered swelling of the crater area.

Between repeated episodes of dome-building, St. Helens steams quietly. Judging by its previous eruptive behavior, the volcano will remain intermittently active for many decades. An eruptive period that began in A. D. 1480 lasted for 150 years, while its nineteenth century activity spanned more than half a century.

St. Helens has produced explosive eruptions far more violent than that of May 18, 1980. About 3,500 years ago it erupted immense quantities of pumice that mantled large areas of the Pacific Northwest with yellowish ash, as far northeast as Banff, Alberta. These outbursts discharged about 2.5 cubic miles of fresh magma, ten times more than was ejected on May 18, 1980. About five hundred years ago the volcano produced two widespread ashfalls, both much larger than that of 1980. St. Helens is capable of producing an explosive eruption catastrophically greater than any event of historic time.

St. Helens' last two eruptive cycles suggest what may happen next. After the violently explosive 1480s, the following decades saw intense cone-building activity. Thick flows of andesite lava alternated with the ejection of pumice and other tephra, rock fragments blown into the air. The cycle ended in the early 1600s with eruption of an immense lava dome that formed St. Helens' pre-1980 summit. The volcano then slept until 1800, when an explosive eruption produced about the same volume of material as the 1980 event. Lava flows were followed by the extrusion of a new lava dome, Goat Rocks, destroyed along with the former summit on May 18, 1980. Sporadic ejections of ash continued until 1857. If the past is a reliable guide, we can probably expect St. Helens to rumble and blaze well into the twenty-first century.

Chapter 14

THE "OTHER" CASCADE VOLCANOES

After the catastrophic eruptions of 1980, St. Helens became a household name, as well as a new national monument. Many Americans do not yet realize that St. Helens is only one of a large family of volcanic siblings, many of which are potentially as destructive as she. The Cascade Range, which stretches from northern California into southwestern British Columbia, contains at least twenty large volcanoes that have erupted violently since the end of the Ice Age, about 10,000 years ago. Several volcanoes, including St. Helens, Glacier Peak, and Mount Mazama (Crater Lake), produced explosive outbursts of such cataclysmic proportions that, if repeated today, they would create a regional disaster.

During the brief span of historic time, roughly the last 250 years, as many as eleven of the Cascade volcanoes have erupted. A century ago few people lived in the areas affected by historic eruptions; today thousands live or vacation in these areas, making hazards from future eruptions a serious social issue.

The prospect of more Cascade eruptions delights volcano-watchers, but worries people who live nearby. Understanding how the volcanoes have behaved in the recent geologic past can suggest ways of coping with their future activity.

Of the eleven historically active peaks, the six most likely to endanger lives and property when they next erupt include Baker, Glacier Peak, Rainier, and St. Helens in Washington state, Hood in Oregon, and Shasta in California. Other potentially explosive volcanoes, such as those in Lassen Volcanic National Park, central Oregon's South Sister and Newberry, and the Medicine Lake volcano in

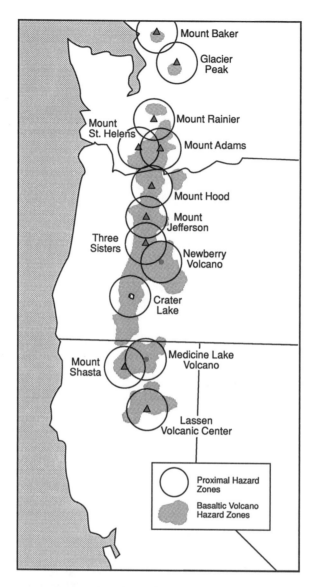

Volcanic hazard map of the Cascade Range. The circled proximal-hazard zones have a thirty-one mile radius encompassing each of the thirteen major potentially explosive volcanic centers. The circled areas are subject to heavy ashfall from future eruptions and / or damage from lava flows, pyroclastic flows and surges, directed blasts, debris avalanches, mudflows, and floods. Stream valleys draining the high glaciated peaks may be inundated by mudflows and floods several tens of miles beyond the circled area. The lighter pink zones represent areas in which less explosive basaltic or basaltic-andesite eruptions may occur. —After Hoblitt, Miller, and Scott, 1987

northeastern California, are located in sparsely populated areas and are not expected to threaten major settlements.

This ranking of potential hazards, however, assumes that the volcanoes' eruptive conduct will resemble that of the last few thousand years. An eruption like the one that formed Crater Lake about 6,900 years ago, the largest explosive event in North America since the close of the Ice Age, would threaten people and property many tens of miles from the volcano.

All six of the most dangerous Cascade peaks are high steep composite cones, built of lava flows and fragmental rock. Their eruptive behavior apparently alternates between relatively quiet emission of fluid lava and violent discharge of rock fragments, such as pumice. No two erupt in exactly the same way. Each has its own distinctive behavior pattern, which commonly includes a wide range of variations.

Despite their diverse individual personalities, all six high risk volcanoes are a genuine threat to any settlement immediately down-valley from their summits. All six are crowned with active glaciers, thick streams of ice that flow down the volcanoes' flanks and commonly terminate above deep canyons and valleys that head on the volcanic cone.

Rainier (elevation 14,410 feet) is mantled with 156 billion cubic feet of ice, more than any other U.S. peak south of Alaska. Baker, Glacier Peak, and Hood also support large glaciers positioned above adjacent valleys in which many thousands of people now live. During previous eruptions at all these volcanoes, the sudden melting of ice and snow generated large floods and mudflows that swept down valleys heading on the mountain and buried valley floors under thick accumulations of volcanic rubble. A turbulent mixture of water and rock debris of all sizes, volcanic mudflows typically move downslope at speeds of twenty to forty miles per hour.

The largest known Cascade muflow, the Osceola, originated high on Rainier's east side about 5,700 years ago and traveled sixty-five miles down the White River valley to inundate at least 125 square miles of the Puget lowland. Its half cubic mile engulfed at least one Indian camp and buried the sites of several Washington towns, including Enumclaw, Auburn, Sumner, and other places where 60,000 people now live. The Electron mudflow of about 600 years ago, which may have been triggered by a great earthquake on the Cascadia subduction zone, flowed west over the site of Orting and other valley settlements.

Mount Rainier supports a larger burden of active glaciers than any other U.S. peak south of Alaska. Heat and steam emission in the bowl-shaped summit crater has melted out a network of caverns and tunnels between the crater walls and icefill, creating an extraordinary ice cave system.
—U.S. Geological Survey photo

Television cameras filming the 1980 floods and mudflows traveling from St. Helens down the Toutle and Cowlitz rivers and emptying into the Columbia provided vivid images of a phenomenon that accompanies almost every Cascade eruption. Because they travel quickly over long distances and obliterate everything in their paths, mudflows are among the most destructive by-products of an erupting Cascade volcano.

Rainier has erupted at irregular intervals during the last 10,000 years, sometimes with catastrophic results. About 2,000 years ago the last major eruption deposited a thick blanket of pumice east of the volcano and built the lava cone at Rainier's present summit. This episode spawned mudflows that raised the White and Nisqually river valley floors eighty feet above their present levels. Several small eruptions sprinkled pumice fragments over Rainier's eastern slopes between about 1820 and 1854.

Second only to Rainier in its extensive glacier cover, Baker has produced numerous avalanches and mudflows during the last several hundred years. Future avalanches from the unstable rock along the east rim of the crater and melting of the Boulder Glacier on Baker's east flank could send a large mudflow into Baker Lake, the major reservoir at the volcano's eastern base, possibly overtopping the dam at high water and generating floods down the Baker River. Although Baker has erupted lava flows during the last 8,700 years, it is more likely to produce moderate quantities of ash and rock fragments that will blanket areas downwind.

Baker holds the Cascade record for the largest number of observed eruptions over the longest period of time. Between 1792, when a Spanish ship's captain recorded a noisy outburst, and 1880, when settlers saw a cloud of steam and ash, witnesses reported at least two dozen eruptive events. The eruption of 1843, which scattered ash over a wide area and spread debris down Baker's east side, was apparently the most severe. A comparable event today could endanger campers and boaters on Baker Lake.

Glacier Peak, southeast of Baker near the crest of the rugged North Cascades, is little known but potentially dangerous. This ice-sheathed cone has repeatedly ejected very large volumes of pumice and produced massive avalanches of hot rock that sent floods and mudflows streaming downvalley as far west as the shores of Puget Sound. Despite its position far from major settlements, the volcano's numerous mudflows have devastated several different valley floors seventy

Heavily sheathed in glacial ice, Mount Baker steams visibly. Between 1843 and 1880 Baker produced numerous small ash eruptions, several of which coincided with St. Helens' historic activity. —U.S. Geological Survey, Robert Krimmel photo

Mount Hood from the north, showing Eliot Glacier. The highest and most recently active volcano in Oregon, Hood poses a threat to people and property located downvalley from its ice-covered slopes. It erupted most recently during the middle of the nineteenth century. —U.S. Geological Survey, Austin Post photo

to one hundred miles away from the cone, including the sites of towns in the increasingly populated Skagit River valley.

Oregon's highest and most frequently active volcano, Mount Hood (elevation 11,245 feet) has staged at least three major eruptive cycles during the last 1,700 years. The earliest and largest created the massive apron of rock debris, about three-quarters of a cubic mile in volume, that makes Hood's smooth south flank so attractive to skiers. During all three eruptive cycles, masses of thick lava oozed from vents just south of the summit ridge, building a prominent lava dome, Crater Rock, that still steams vigorously.

As the tongues of sticky lava extruded, they crumbled and partially collapsed, generating avalanches of gas-charged hot rock, pyroclastic flows, that swept downslope into the Sandy and White river drainages. Melting icefields poured mudflows into the Sandy, Zigzag, and White river valleys, some of which reached the Columbia River. The latest eruptions, between A.D. 1760 and 1810, sent floods and mudflows down the Tygh Valley and the Sandy River to its confluence with the Columbia, burying the sites of several modern towns and resorts.

After a few decades' repose, Hood reawakened in 1859 and 1865-66, producing a series of mildly explosive eruptions, one or more of which scattered light gray pumice fragments over the mountain. If future activity follows the pattern established during the last 1,700 years, lava erupted from vents behind Crater Rock will direct pyroclastic flows and extensive mudflows into the White River drainage on the volcano's southeast side and into the Sandy and Zigzag river drainages to the west.

California's potentially most dangerous volcano is also the state's most magnificent mountain, a cluster of four overlapping cones visible for one hundred miles in every direction. Towering nearly 11,000 feet above Interstate Highway 5, which crosses its western base, Mount Shasta has erupted often and violently during the last few thousand years.

Like Mount Hood, Shasta typically erupts masses of thick, gooey lava at or near the summit. Collapsing or shattered by steam explosions as they rise, Shasta's lava domes commonly form large pyroclastic flows derived from incandescent dome fragments. Traveling at

Mt. Hood volcanic hazards map. Mt. Hood erupted repeatedly between about A.D. 1760 and 1810, triggering a series of floods and mudflows into the Sandy, Zigzag, and White river drainage basins. Thick mudflow deposits mantled Old Maid Flat and traveled down the Sandy River as far as Marmot Dam; more dilute runout poured into the Columbia, temporarily forming a sandbar across the river. Avalanches of hot rock erupted into the upper White River valley generated extensive mudflows that extended through Tygh Valley into the Deschutes River. Hood produced smaller eruptions in 1859 and 1865-1866. —Modified from Cameron and Pringle, 1987

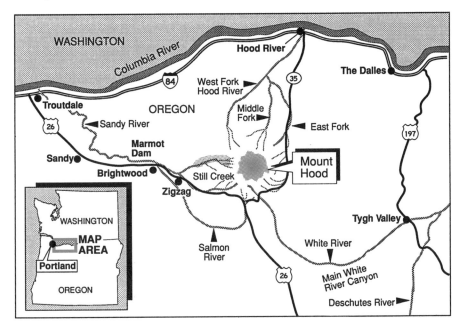

high speeds, pyroclastic flows have rushed down every side of Shasta's vast edifice, incinerating and burying everything in their paths.

The towns of Weed and Mount Shasta stand on pyroclastic flow deposits erupted from the summit of Shastina, the massive cone on Shasta's western flank, or from Black Butte, a cluster of lava domes of about the same age. McCloud, the picturesque village at Shasta's south base, stands on recent mudflow deposits. These settlements, as well as Dunsmuir, are huddled at the base of a sleeping giant that may become a lethal neighbor.

Shasta had its last major eruption in 1786, when the explorer La Perouse, then sailing off the northern California coast, saw it in action. That activity produced hot and cold mudflows and a large pyroclastic flow that swept from the summit crater more than seven miles down Ash Creek on the volcano's east flank. When Shasta revives, new vents may open high on the mountain's west side, possibly directing pyroclastic flows down the same routes they have traveled before, plowing through Weed and Mount Shasta, instantly carbonizing all structures and their inhabitants.

Large eruptions may also occur at other recently active Cascade volcanoes, including several in northern California. Oregon's Newberry and South Sister seem to be evolving a magma capable of producing a world-class explosive eruption comparable to the one that leveled ancient Mount Mazama and formed Crater Lake (Chapter 27). All historic eruptions in the forty-eight adjacent states have occurred in the Cascade Range, a reminder to the three million people living within sight of them that these glacier-wrapped peaks represent a terrible beauty.

Chapter 15

SCENERY AND SOLITUDE:
Lassen Peak and the
Medicine Lake Volcano

Competing for breathing space with hordes of tourists can take the fun out of summer travel. Americans seeking to enjoy mountain scenery free from the crowds that flock to Yellowstone or Yosemite can find quieter ground in northern California's Lassen Volcanic National Park and Lava Beds National Monument.

Lassen Park recently observed the seventieth anniversary of its creation (1916)—appropriately celebrated by maintaining its traditionally low quota of visitors. Vacationers soon discover that the Park offers both a rugged forest landscape dotted with emerald lakes and a spectacular window into the earth's hot interior.

Lassen Park has hot springs, roaring steam vents, bubbling mud pots, and three young volcanoes that have erupted during historic time. Until St. Helens awoke in 1980, Lassen Peak (elevation 10,457 feet) was advertised as the most recently active volcano in the continental United States. After it revived in May, 1914, Lassen produced at least 170 explosive eruptions during its first year out of retirement. The climax came on May 22, 1915, when an immense mushroom cloud of steam and ash shot an estimated seven miles into the air, showering fine grit over Nevada communities 200 miles to the east. Simultaneously, a seething mixture of hot gas and incandescent rock fragments swept down the volcano's northeastern flank, obliterating a forest and scouring the ground clean up to four and half miles from the summit.

The 1915 eruptions also generated a lava flow and several large mudflows that devastated the upper twenty miles of Lost and Hat creeks. After blasting open a new crater in 1917, Lassen's activity gradually declined. Except for wisps of steam leaking from crevices in the summit craters, it has been silent since 1921.

Despite the main peak's tranquility, Lassen Park is rich in thermal activity. The most vigorous thermal displays are associated with a much older volcano called Mount Brokeoff, or Mount Tehama. Once a towering giant, Brokeoff is now deeply eroded, its former lofty summit replaced by a deep depression that houses clusters of steaming fumaroles. Route 89, the Park's main road, crosses Brokeoff's glacier-carved interior, bringing the visitor to the roadside Sulphur Works, suitably named for the strong rotten egg odor of hydrogen sulfide issuing from dozens of steam vents and hot springs.

Lassen Peak during a minor explosive eruption, October 6, 1915, viewed from Reflection Lake, several miles west of the volcano. After the climatic outburst of May 22, 1915, Lassen continued to produce eruptions of varying intensity through June, 1917, when a new crater was blasted open on the northwest side of the summit. Large plumes of steam rose intermittently from the crater until February, 1921. —Lassen Volcanic National Park, Chester Mullen photo

One can stroll through the hissing Sulphur Works in a few minutes. A longer hike takes one through the nearby inferno called Bumpass' Hell. Scores of vents shoot columns of steam skyward. Hot springs boil fiercely, miniature volcanoes with craters full of liquid mud cough globs of hot mud that build miniature cones that mimic those of great volcanoes.

The ground surrounding these sputtering craters and fumaroles is stained brilliant shades of orange, rust, yellow, and white, the colors of minerals that form as solid rock decays into opal and clay. Steam, hot water, and acidic gases are transforming Brokeoff's core into soft and easily eroded clay.

A well-graded trail leads to Lassen's summit, from which one can enjoy a bird's eye view of the Park's many volcanic features. Immediately north of Lassen are the Chaos Crags, a cluster of lava domes erupted about 1,100 years ago. About 300 years ago part of the northern dome collapsed, triggering a massive rockfall and avalanche, the Chaos Jumbles.

Cinder Cone, in the Park's northeast corner, erupted as recently as 1850-51, scattering gray ash over a broad area. Cinder Cone and other volcanoes in the eastern part of the Park erupt basaltic magma that is typically gas-poor and non-explosive. A relatively shallow body of gas-rich dacite magma has evolved under the western edge of the Park and produces violently explosive eruptions. Supplying Lassen Peak, Chaos Crags, and other near-by vents with highly volatile fuel, this magma reservoir may generate catastrophic eruptions in the future.

Lava Beds National Monument, more than 100 miles northeast of Lassen Park, offers its few visitors a spectacular variety of volcanic landforms. The Monument's surface consists largely of three voluminous basaltic lava flows riddled with miles of lava tubes, caves and tunnels formed when the lava was still hot and flowing. Of the 300 known caves, more than twenty have been developed for public use.

The Lava Beds tubes strikingly resemble similar formations in Hawaii. Such tubes typically form in pahoehoe lava, a Hawaiian term used to describe a fluid basalt flow with a smooth, ropy surface. Erupted at extremely high temperatures, commonly about 2,000 degrees F., pahoehoe lava moves forward as the molten rock of the flow's interior drains away from beneath the solidifying surface crust. Draining through tubes inside the lava stream, the liquid material advances the front of the flow and leaves a hollow flow channel behind when the eruption ceases. These lava tunnels become accessible only after the thin crust roofing the tube collapses, creating an opening to the flow interior.

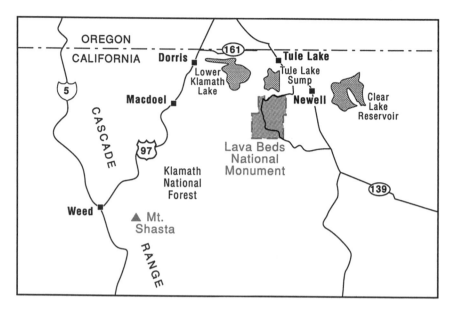

Lava Beds National Monument lies south of Tule Lake at the northern foot of the sprawling Medicine Lake volcano, largest in the Cascades. During recent eruptions, fluid streams of basalt covered the volcano's flanks while violent explosions occurred at numerous vents near the summit.

The Lava Beds are part of the Medicine Lake volcano, one of the largest volcanic structures in the United States. Although it stands only about 4,000 feet above its base, the volcano sprawls over approximately 900 square miles and has an estimated volume of 130 cubic miles.

Although tucked away in the sparsely populated northeastern corner of California, the Medicine Lake volcano may soon shed its obscurity. During late September and early October, 1988, a series of small earthquakes centered beneath the summit caldera, a shallow depression holding Medicine Lake. The quake swarm has since died down, but it reminds us that molten rock stirs under this recently active volcano.

Medicine Lake's outer slopes are covered with basaltic lava that normally flows quietly from vents on the volcano's flanks. The latest summit eruptions, however, were extremely violent and spewed large volumes of gas and frothy volcanic glass high into the air, blanketing a large area with pumice as far west as Mount Shasta, thirty-five miles distant.

Looking west from Glass Mountain, a massive rhyolite lava flow erupted atop the Medicine Lake volcano less than 1,000 years ago. Pumice from this eruption blanketed Mount Shasta, thirty-five miles distant. —California Bureau of Mines and Geology, Mary Hill photo

After the explosive phase of the eruption, thick viscous masses of rhyolite lava surged from a chain of at least thirteen vents on the eastern rim of the summit caldera. The lava chilled quickly to form the glistening black glass, obsidian, that forms Glass Mountain. On the western caldera rim, another obsidian flow created Little Glass Mountain.

Future eruptions of the Medicine Lake volcano will probably replicate these recent eruptions, which took place less than 1,000 years ago. Earthquakes strong enough to fracture the ground surface accompanied a small ash eruption in 1910. The 1988 seismic activity may not lead to another eruption, but it could be the first warning that a slumbering giant is about to awake.

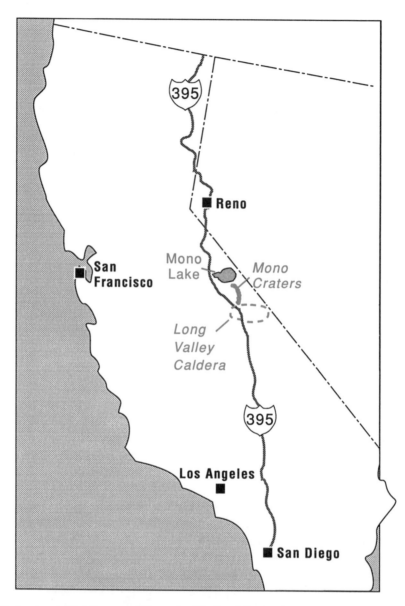

Lying immediately east of Yosemite National Park, the volcanic Mono Lake-Long Valley region marks the boundary between two great geological Provinces, the Sierra Nevada mountains and the Basin and Range province. The westward-drifting crust of the continental interior collides with the massive uplifted granite blocks of the Sierra, generating numerous earthquakes and volcanic eruptions in the process.

Chapter 16

WHAT'S BREWING NEAR MONO LAKE?

Mark Twain described it as "the loneliest spot on earth...a lifeless, treeless, hideous desert." Dramatically bleak, the landscape around Mono Lake in east-central California has changed little since Twain visited it in the 1860s.

Arid and windswept as ever, the only new element in this thinly populated region is the attention it now draws from the news media. National concern focused on the area in May of 1980 when a flurry of earthquakes, registering up to 6.0 on the Richter scale, sent boulders crashing down from the nearby Sierra Nevada and cracked walls in the resort community of Mammoth Lakes. The equally strong quakes of mid-July 1986, although centered near the White Mountain fault zone several miles to the east, continue to excite interest because this unstable area has a potential for geologic violence on a cataclysmic scale.

After St. Helens in Washington state, the U.S. Geological Survey regards the Mono Lake area, with the Long Valley region to the south, as the most likely location of the next major volcanic eruption in the forty-eight adjacent states. Explosive eruptions comparable in size to that of St. Helens in May, 1980, have occurred repeatedly along a chain of volcanic vents extending southward from Mono Lake to the Mammoth Lakes basin. The Mono Craters, a rugged line of cinder cones, craggy obsidian domes, and steep-sided lava flows, are so geologically youthful that they look as if they grew only yesterday. The Inyo Craters, a southward extension of the chain, contain a similar accumulation of jagged lava formations, including spectacular funnel-shaped explosion pits, some of which hold turgid green pools.

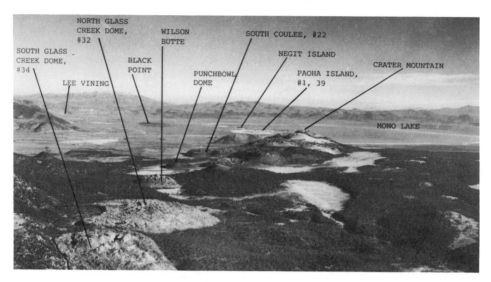

NORTH GLASS
CREEK DOME,
#32

WILSON
BUTTE

SOUTH COULEE, #22

SOUTH GLASS
CREEK DOME,
#34

BLACK
POINT

NEGIT ISLAND

CRATER MOUNTAIN

LEE VINING

PUNCHBOWL
DOME

PAOHA ISLAND,
#1, 39

MONO LAKE

A long chain of thick lava flows and cones, the Mono-Inyo craters, extends southward from Mono Lake, east-central California. Repeatedly active during the last 2,000 years, the craters deposited thick layers of ash over a wide area in western Nevada and eastern California during the mid-14th century. Smaller earthquakes that rattle the Mono Lake-Long Valley region suggest that molten rock continues to move toward the surface. —California Division of Mines and Geology photo

Geologists have long known that the Mono-Inyo Craters have been recently active, erupting from at least thirty different vents during the last 2,000 years, but only lately have the nature and extent of the activity been recognized. Two important new studies by Dan Miller of the U.S. Geological Survey and Kerry Seih of the California Institute of Technology reveal that volcanic action at the Mono-Inyo Craters differs significantly from that at most well-known volcanoes.

Famous volcanoes like Vesuvius, Augustine, or St. Helens are single cones that normally erupted from a single central vent. When the Mono area volcanoes become active, they simultaneously open a whole series of separate new vents along a zone several miles long. The eruptive effects are thus not confined to the vicinity of an individual cone but extend over a wide territory. When the Mono-Inyo Craters last erupted, they spewed lava from a dozen different vents at the northern and southern ends of the chain.

Geologists believe that past events provide some clue to what will happen in the future. If so, then people now living along Highway 395, the main route connecting northern and southern California east of the Sierra crest, can expect a replay of the chain of fire that blanketed thousands of square miles with volcanic debris when the Mono-Inyo Craters last erupted. By counting annual growth-rings on trees rooted in the volcanic deposits, researchers conclude that two spectacular series of eruptions happened between about A.D. 1325 and A.D. 1365 during a very brief period of time, perhaps a few weeks or months.

Activity began at the northern end of Mono Craters when a dike, a thin, vertical body of molten rock injecting a fracture, reached the surface. As it encountered ground water that flashed into steam, the magma exploded with tremendous force, blasting a four mile-long line of new vents and propelling columns of hot ash miles into the stratosphere. The ash clouds, ejected in a cycle of distinct pulses, drifted many tens of miles north, east, and south on changing winds, depositing a thick layer of ash over central California and western Nevada. If a comparable eruption were to happen today, it would produce a layer of ash eight inches thick twenty miles downwind, and as much as two inches thick fifty miles away from the volcano.

Following the vertical expulsion of ash, a second phase of the activity produced pyroclastic flows, turbulent mixtures of seething gas and incandescent rock fragments that pour over the ground surface like a heavy liquid. Unlike the widespread ashfalls, the pyroclastic flows traveled no farther than half a mile from the source vents.

The third and final stage of the eruptive cycle was much quieter. Thick tongues of viscous lava squeezed from several of the vents, like putty from a tube. Too stiff and pasty to flow far, the lava formed short, stumpy lava flows called coulees, and massive domes—protrusions of rhyolite lava that plugged and capped the vents from which they emerged. Five separate domes and coulees were emplaced, the most conspicuous of which is Panum Dome, the prominent lava plug standing near the south shore of Mono Lake.

Impressive as the North Mono eruptions were, they were only the first act in a two-part volcanic drama. Perhaps only a year or two, or less, after the North Mono ash had fallen, another series of explosive outbursts began at the Inyo Craters, a few miles to the south. The sequence of events was similarly devastating to the surrounding countryside. Again, a dike of rhyolitic magma, about four and one half to eight miles long, rose to within about 700 feet of the surface before erupting. A chain of new craters was ripped open, generating towering plumes of pulverized rock that laid down a thick blanket of gray ash, mantling even regions many miles to the west. While some of the new vents spewed fresh magma, others produced eruptions of steam that hurled old rock fragments high into the air.

After the explosive phase, the degassed magma surged into several vents along the Inyo Craters chain, forming steep-sided domes or flows of glistening black obsidian at the South Deadman, Glass Creek, and Obsidian Flow vents.

The Mono-Inyo Craters have been quiet since their double-barreled display about six hundred years ago. Most geologists would not be surprised to see activity resume at any time. During the last two millennia, eruptions have occurred in the area every few centuries, and we may be overdue for another violent episode when the earth's surface will split along fractures several miles long and spew fountains of fire. Whether the 1980s earthquake sequence is triggered by the subterranean intrusion of a molten dike that is destined to erupt at the surface as its predecessors did, only time can tell.

Chapter 17

THE COLUMBIA RIVER PLATEAU: CHAOS FROM HEAVEN

If no man is an island unto himself, impervious to external influences, neither is our planet. One of the most exciting new scientific proposals is that earth's geological development has been significantly determined by extraterrestrial objects. Large meteorites striking the earth may have produced not only mass extinctions of plant and animal life, but also have initiated plate movement and cataclysmic volcanic eruptions.

According to three geologists at the University of Montana, about 17 million years ago a large meteorite struck near the southeastern corner of Oregon. The colossal impact triggered a series of geologic processes that radically transformed the West's topography. With its crust fractured from the impact, the earth bled molten rock as if from an open wound. Immense floods of basaltic lava poured from crustal fractures tens of miles long, inundating vast areas of eastern Oregon, Washington, western Idaho and northern California.

Highly fluid streams of basalt lava traveled as far as 300 miles from their source and spread out in broad sheets that buried the previously existing landscape. No eruptions in historic time have come near to equaling the enormous volumes of molten rock poured out to create the Columbia River Plateau, by far the largest volcanic landform in North America. Spreading over the Northwest's interior like a fiery tidal wave, the basaltic lava traveled west through the Columbia River's ancient channel as far as the Pacific. Some lava flows filled valleys and other depressions between individual peaks of the ancestral Cascade Range.

POSSIBLE CLIMATIC EFFECTS OF BASALTIC FLOOD ERUPTIONS

The enormous floods of basaltic lava that rolled west from eastern Washington to the Pacific buried tens of thousands of square miles during a single eruption. Fire fountaining along extensive fissures and heat and gas escaping from a vast sea of molten rock undoubtedly affected earth's climate. Convective plumes would carry masses of sulfur-rich gases, the precursors of sulfuric acid aerosols, high into the stratosphere, where winds would transport them around the hemisphere or even the globe.

The largest basaltic eruptions of historic time, the 1783 eruptions of Iceland's Laki volcano, showed how severely this kind of lava outpouring can afflict the environment. Fissures eighteen miles long discharged fluid lava at the rate of 5,000 cubic yards per second and buried about 340 square miles. Flooding from melted ice and lava-blocked rivers destroyed much property, but the eruption's worst effects stemmed from a blue haze, probably containing hydrogen sulfide and other gases, that discolored Iceland's atmosphere for months. Retarding plant growth, the haze caused crop failures and a disastrous famine. Iceland lost a fifth of its population, about two thirds of its sheep and horses, and half of its cattle.

Although it produced only a tiny fraction of the lava and gases erupted on the Columbia Plateau, the Laki eruption, along with an unusually large ash eruption of Japan's Asama volcano the same year, significantly affected global climate. Benjamin Franklin noted that a dry fog persisted over Europe and much of North America the summer of 1783 and that the following winter was unusually cold. The poor growing season and resulting economic losses helped bankrupt France and precipitate the French Revolution.

In some areas, ten thousand years passed between eruptions, allowing reddish soils to develop on lava flow surfaces before they were covered by new effusions of molten rock. In terms of geologic time, however, the Columbia Plateau lavas erupted with astonishing rapidity. Most of the estimated 60,000 cubic miles of basalt poured out during a relatively brief time span, between about 17 million and 15 million years ago. The peak of effusive activity lasted only about a half million years.

The Columbia River Plateau, covering more than 200,000 square miles, is the earthly counterpart of a lunar mare, an immense lava "sea" formed on the moon's surface following a meteorite impact. An even larger terrestrial mare is the Deccan Plateau in western India, an enormous impact crater that filled with basaltic magma from the

According to the meteorite impact theory, the enormous outpouring of basaltic lava that began about 17 million years ago resulted from a meteorite striking southeastern Oregon near the Idaho border. The lava flows composing the Northern Columbia Plateau in eastern Washington were generated by a northward extension of crustal fractures from the impact site. As North America traveled westward over the impact area, which formed a persistent hot spot, eruptive activity migrated northeastward along the Snake River Plain to its present center beneath Yellowstone National Park. —After Alt and others, 1989

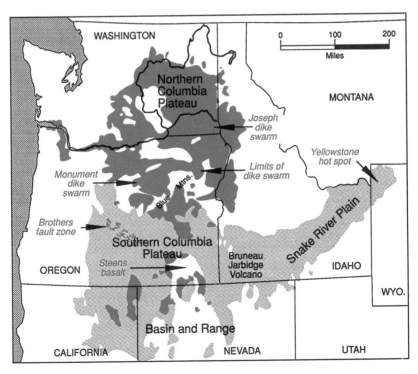

Sequence of cataclysmic events leading to the creation of the Columbia River Plateau, the Snake River Plain, and the Yellowstone volcano.

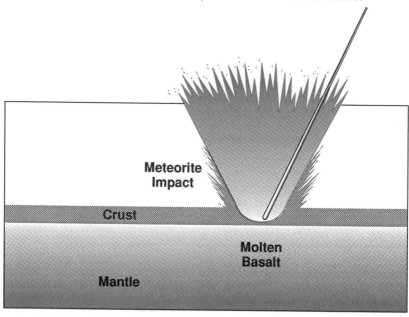

A. About 17 million years ago a large meteorite strikes near the present Oregon-Idaho boundary, destroying crustal rock at the impact site. With the confining pressure of overlying rock removed, hot plastic material in the mantle expands and rises to the surface, forming a giant sea of liquid rock.

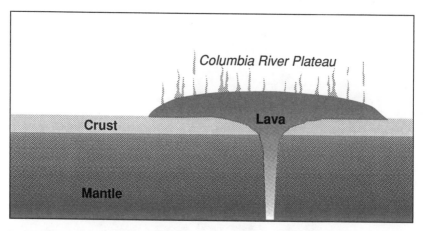

B. Molten rock rising from the upper mantle floods the impact site and overflows in enormous sheets of fluid basaltic lava. Single streams of lava travel up to 300 miles westward to the Pacific. Repeated basaltic flood eruptions build the Columbia River Plateau of eastern Oregon, Washington, and southwestern Idaho.

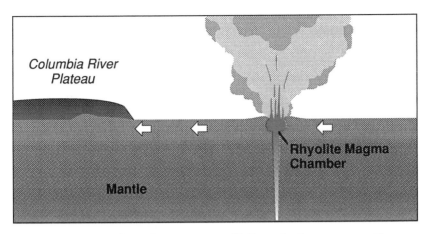

C. A hot spot, a persistent low-pressure cell, forms in the upper mantle, through which heat constantly rises to the surface. As the North American plate moves westward, the voluminous outpourings of basaltic lava gradually diminish at the Columbia River Plateau and volcanic activity shifts to the east. When undamaged granitic crust moves across the hot spot, it partly melts, producing subterranean reservoirs of water-saturated rhyolitic magma that erupts violently in huge clouds of incandescent ash.

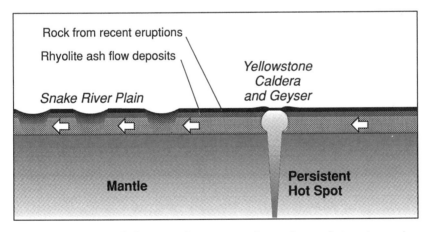

D. As the continental plate travels west over the stationary hot spot, a series of rhyolitic ash flow eruptions form a sequence of calderas, volcanic collapse depressions, across the Snake River Plain of southern Idaho. The hot spot now simmers beneath the Yellowstone volcano, site of three catastrophic outbursts during the last 1.2 million years. More recent eruptions of basaltic lava, such as those at the Craters of the Moon National Monument, form a thin veneer of dark rock over the light-colored rhyolitic ash flow deposits.

131

View south across the Columbia River at Biggs Junction, Washington. The nearly horizontal ledges of basalt outline basalt lava flows. The level distant horizon in Oregon is the original volcanic surface of the Columbia Plateau. —Donald Hyndman photo

earth's mantle and overflowed to create the world's largest lava fields. The meteorite that exploded on impact to initiate the Deccan Plateau eruptions about 65 million years ago may also have been the source of the iridium deposits that mark the end of the Cretaceous period. The discovery of widespread iridium deposits, an element rare on earth but more abundant in meteorites, supports the theory that collision with an extraterrestrial object led to the catastrophic extinction of species that closed the Cretaceous period. Global fallout from the impact, plus the long-continuing volcanic eruptions it triggered, help to explain the rapid disappearance of most large animals, including the dinosaurs.

The effects of a large meteorite impact do not end with the creation of a voluminous lava sea at or near the collision site. The impact shatters the earth's lithosphere, the rigid rocky crust and plastic upper mantle, allowing hot material of the interior to expand, liquify and erupt on the surface. By removing the containing weight of overlying rock, the impact will probably create a persistent low pressure cell below the crust, a mantle plume or hot spot through which heat rises continuously to the surface.

A hot spot remains permanently rooted in the mantle, but the earth's outer shell, broken into rocky slabs or plates, continues to move. As the plate carrying North America on its back traveled slowly westward, eruptions gradually ceased in the Columbia Plateau region. As southern Idaho slid over the hot spot (impact site), violently explosive eruptions occurred, forming a series of huge calderas, each of which is east of its predecessors. The last of these great explosive outbursts ejected huge quantities of rhyolite ash that devastated thousands of square miles about 600,000 years ago and opened the vast Yellowstone caldera. Today the hot spot lies beneath Yellowstone National Park, providing the thermal energy for the park's spectacular geysers, fumaroles, and hot springs.

The Snake River Plain, extending east-northeastward from the southern Columbia Plateau in Oregon, across southern Idaho, to its apex at the Yellowstone caldera in northwest Wyoming, marks the track of the continent's westward migration over the hot spot. As the continental crust passed over the hot spot, it partially melted to create a series of huge volcanoes, of which the Yellowstone volcano is the youngest. Eruptions began at the southwest corner of the Snake River Plain about 13 million years ago and have continued intermittently at its northeastern tip almost until the present. The hot spot may eventually burn northeast to North Dakota, perhaps beyond.

Columns of Columbia River basalt along U.S. Highway 395, south of the Washington-Oregon border.
—Donald Hyndman photo

Besides creating the vast Columbia River lava plateau, the hot spot that generated the Snake River Plain, and eventually the Yellowstone caldera, the meteorite impact apparently also initiated the continental rifting, or splitting apart of the earth's crust, that formed the northern Basin and Range region of western Utah, Nevada and northeastern California. The Basin and Range Province, which includes much of the western interior, is characterized by alternating mountain ranges separated by broad depressions or basins. This rugged terrain is thought to be caused by the extension and thinning of the crust as North America moves westward. As the crust thins and fractures through extension it breaks into crustal blocks, some of which sink to form basins while neighboring blocks rise to form mountain ranges. The crustal fracturing and spreading that formed the northern Basin and Range region began about 17 million years ago, precisely when the basaltic flood eruptions began at the presumed site of impact in southeastern Oregon. While the Snake River eruptions progressed steadily eastward as the continent passed over the hot spot, so the northern Basin and Range area also expanded eastward as it traveled over the spreading zone created by the impact.

Today most volcanic activity takes place along linear zones where earth's tectonic plates pull apart or collide, particularly along continental margins where heavier sea floors are sinking or being subducted beneath lighter continental land masses. It has long been a mystery why some of the greatest eruptions of the geologic past—such as the basaltic floods that formed the Deccan and Columbia River lava plateaus—occurred in continental interiors, far from subduction zones. The new theory provides a comprehensive model that explains the formation of our vast lava seas, the origin of hot spots, and even the trigger mechanisms of intra-continental rifting. It also demonstrates that the earth is not a closed system that determines its own destiny, but that our planet's geology—and life history—may be shaped by foreign objects arriving from distant reaches of the solar system.

Cross section of a lava blister north of Twin Falls, Idaho. —Donald Hyndman photo

Chapter 18

THE SNAKE RIVER PLAIN:
Tracking a Geologic Hot Spot

To most travelers, the vast flatlands of eastern Washington, Oregon, and southern Idaho look bleakly alike. Arid and typically almost featureless, the black lava wastes stretch monotonously between the Cascade Range and the Rockies, occupying approximately 200,000 square miles of the Pacific Northwest's interior.

To have visited one of the West's immense lava fields, however, is not to have seen them all. Despite its superficial resemblances to the basaltic plateaus of Washington and Oregon, Idaho's Snake River Plain is really quite different. Its basaltic surface looks like those in the Columbia River region, but the similarity is only skin deep. Beneath the plain's thin façade of dark basalt lie much thicker layers of light-colored volcanic rock, rhyolite, that was deposited during tremendously explosive eruptions of ash. The quietly erupted lava streams forming the present surface mask vast deposits left by repeated episodes of catastrophic violence.

The Snake River Plain's unusual configuration, aerial extent, and location offer clues to its origin. Shaped like a crescent moon lying on its back, the plain curves across southern Idaho, with its northeastern tip extending almost to the Wyoming border. Surrounded by towering mountains for most of its length and heading at the active Yellowstone volcano, the plain marks the path of North America as it passed over a stationary hot spot in the earth's mantle. The hot spot originated suddenly approximately 17 million years ago, probably the result of a large meteorite impact near the southeastern Oregon-Idaho border.

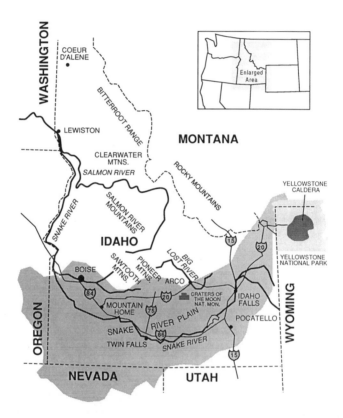

Idaho's Snake River Plain formed as the North American continental plate traveled west across the site of a hot spot in the mantle that now lies beneath Yellowstone National Park. The hot spot partially melted crustal rock moving above it to generate a series of violently explosive eruptions, each to the east of its predecessor. A repeat of the most recent eruptions, which formed the Yellowstone caldera, would devastate much of the north-central United States.

The presumed meteorite striking southeastern Oregon opened a gigantic crater in the earth's crust, allowing the hot plastic rock in the upper mantle to liquify and flood the impact site with basaltic magma. After enormous outpourings of basalt lava had created the Columbia River Plateau, a strikingly different kind of eruptive process began in southwestern Idaho about 13 million years ago. The meteorite had presumably destroyed the continental crust at the impact site, but when the undamaged granitic crust to the east moved over the hot spot a different type of magma and mode of eruption were generated. As superheated basalt from the mantle hot spot rose into the granitic rocks of the crust, it melted them to form rhyolitic magma, which is far more explosive than the gas-poor basalt derived directly from the mantle.

As North America rumbled across the persistent hot spot, extremely large water-saturated pockets of rhyolitic magma developed in the crust. Heat rising from these underground reservoirs of molten rhyolite caused the overlying crust to inflate, buckle, fracture, and collapse, allowing the gas-charged magma to escape explosively. Occurring sporadically during a period of several million years, the rhyolitic eruptions discharged many hundreds of cubic miles of superheated ash that swept through southern Idaho and the surrounding region. Fallout from these glowing clouds deposited thick ash layers east of the Rockies and over the northern plains states.

Each of the major explosive eruptions opened a large caldera, in turn forming a series of deep basins along the Snake River Plain. Some geologists estimate that if all the ash derived from the Idaho calderas were collected together it would approximately fill the long

These bluffs near Hammett, Idaho reveal a thin basalt flow capping grass-covered slopes of rhyolite. —Donald Hyndman photo

trench underlying the Snake River Plain to about the level of the mountains that now surround it. In short, the caldera-forming eruptions literally blew away the previously existing mountainscape.

Although it is largely undissected by erosion, the Snake River Plain has been cut by a few streams to reveal that the basaltic lava surface is merely a thin veneer covering thick deposits of rhyolite. Where exposed by roadcuts, the dark basalt contrasts vividly with the light gray, yellow, or pale pink rhyolite beneath it.

The young basaltic flows that mantle the older rhyolite ashflows were not erupted from the series of calderas and are not the result of hot spot activity. They issued from typically north-south-oriented fissures that probably result from the thinning and stretching of the continental crust as it moves westward at the rate of about two inches a year. As they are stretched, the brittle crustal rocks break and fracture, permitting deep-seated basaltic magma to rise through the fractures and erupt on the surface.

The Great Rift, Great Rift National Landmark, northwest of American Falls, Idaho. —Donald Hyndman photo

Craters of the Moon

The most recent Snake River Plain eruptions centered in southeastern Idaho at the Craters of the Moon National Monument. The latest activity there, which ended scarcely two thousand years ago, was probably typical of that which formed the entire plain surface during the last one million years.

The Craters of the Moon activity resembled that currently taking place in Hawaii, but on a larger scale. The eruptions occurred intermittently along a zone of parallel fractures known as the Great Rift, which extends southeast from Craters of the Moon across the plain almost to American Falls, a distance of about fifty miles.

As at Hawaii's Mauna Loa and Kilauea, fountains of incandescent basalt played along linear fractures, feeding immense rivers of molten rock. Originating at the Great Rift, lava flows traveled as far as twenty-eight miles southeast, and thirteen miles northwest. Altogether, this series of flows buried an area of about 600 square miles. These are large and long flows but not nearly in the same league as the flood basalt flows that built the Columbia Plateau.

When mildly explosive activity persisted at a given location along the Rift, steep-sided cinder cones were erected. Dozens of these cones dot the Monument's largely treeless landscape. The largest is Big Cinder Butte, which stands 800 feet above its base. A brisk hike to the summit affords a sweeping panorama of the lunar features that gives the Craters of the Moon their name.

In terms of geologic time, the volcanoes of the eastern Snake River plain spewed lava only yesterday. Earth scientists believe that eruptions similar to those at Craters of the Moon will occur again, perhaps in the same general location and perhaps within the next few centuries. Although spectacular, anticipated eruptions on the Snake River Plain will be incomparably less troublesome than those expected to occur at the nearby Yellowstone caldera.

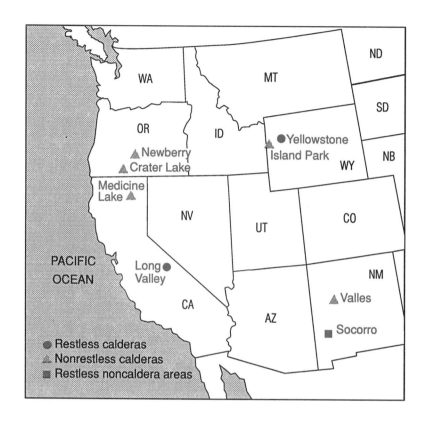

Map showing location of several geologically young calderas in the western United States. Restless calderas, such as those at California's Long Valley and Wyoming's Yellowstone volcano, are characterized by current earthquake activity, uplift of the ground surface by movement of subterranean magma, and emission of steam or other volcanic gases. Presently quiet, the Crater Lake caldera in southern Oregon formed only 6,900 years ago when an explosive eruption destroyed the former summit of a large Cascade volcano, Mount Mazama.

Chapter 19

TORRENTS OF FIRE:
Eruptions That Can
Devastate A Continent

Foaming silver-white against a blue sky, the plume of steam and hot water that Old Faithful regularly shoots into the air enthralls millions of visitors to Yellowstone National Park every year. Despite forest fires and midsummer overcrowding, the park's thousands of boiling hot springs, spouting geysers, and brilliantly colored mineral terraces seldom fail to delight.

Yellowstone's energetic displays of earth's internal heat are awe-inspiring, but they only hint at the stupendous volcanic forces that created them. America's first and most famous national park was the site of three of the most powerfully destructive volcanic eruptions known. It may be evolving toward a fourth.

The rugged Yellowstone Plateau in northwestern Wyoming averages about 8,000 feet above sea level and is flanked by mountains over 12,000 feet high. To the west and southwest the mountainous terrain drops gradually to the eastern Snake River Plain. The Teton Range extends south into Grand Teton National Park, where a chain of spectacularly jagged peaks seems to rise almost vertically above their eastern base in Jackson Hole, Wyoming. Before the Yellowstone eruptions violently rearranged the Rocky Mountain landscape, similarly picturesque peaks may have extended across the Yellowstone park area.

Volcanic activity in the Yellowstone region began about 2.2 million years ago, apparently as this part of the continental crust moved westward across the persistent hot spot that had earlier created the Snake River Plain. Subterranean melting of the crust produced a large body of magma that intermittently leaked to the surface. The earliest eruptions were relatively small, but by two million years ago enough magma had accumulated underground to produce one of the largest explosive eruptions ever to occur on earth. The most voluminous of the three great Yellowstone eruptions, this outburst ejected more than 620 cubic miles of fresh magma. That is enough to build six mountains the size of California's 14,162-foot Mount Shasta—all blown out in a single eruption!

No eruption of comparable size has occurred during historic time, but we can infer what happened at Yellowstone from recent similar but much smaller eruptions. The 1912 outburst that created Alaska's Valley of Ten Thousand Smokes ejected a large volume of rhyolitic magma that swept across the landscape as an incandescent ash flow resembling those erupted by the ancient Yellowstone volcano (Chapter 23).

Suddenly released from the confining pressure of the overlying crustal rock, gases contained in the magma expanded with tremendously explosive force. Escaping hot gas whipped the magma to a glassy froth that burst simultaneously from a series of concentric crustal fractures that had formed above the underground magma reservoir. After jetting to stratospheric heights, a giant column of gas and incandescent magma fragments collapsed to spread outward in all directions from its source. The outward moving waves of seething gas and frothing lava traveled enormous distances, sweeping over ridges and peaks and filling valleys with molten ash. Despite the staggering volume of magma spewed out, the eruption lasted only a few hours or days. The ash flows spread over thousands of square miles so quickly that the incandescent rock lost little heat in transit. Deposited at extremely high temperatures, glassy fragments in the ash flows melted and fused together, forming ignimbrite, literally rocks formed from glowing clouds.

So much material was ejected that the roof of the underground magma reservoir collapsed, causing the overlying crustal rocks to subside several thousand feet. The resulting caldera covered an area of about 1,000 square miles. Extending across Island Park and the Yellowstone Plateau, the caldera outlines have been largely obliterated by later eruptive and erosive activity.

The second volcanic cycle largely duplicated events of the first. The smallest of the three catastrophic episodes, the Island Park cycle was centered inside the east end of the original caldera, near the head of the Snake River Plain. After quietly erupting lava flows of rhyolite and obsidian, about 1.3 million years ago the Island Park volcano produced Yellowstone's second climactic outburst. Disgorging about seventy cubic miles of magma in the form of rhyolitic ash flows, the volcano collapsed, forming the Island Park caldera, which measures about seventeen miles across. Again, the explosive event was followed by the emission of numerous rhyolite lava flows from a chain of vents that stretches northwestward across the caldera. The magma chamber beneath the Island Park caldera has now completely solidified and been broken into crustal fractures, through which basaltic lavas have since erupted.

The third and latest cycle began about 1.2 million years ago and was centered considerably east of Island Park beneath the present Yellowstone Plateau. Rhyolitic lava flows erupted intermittently along a slowly forming set of circular fractures, gradually outlining the dimensions of the future third caldera. The series of concentric ring fractures would serve as eruptive vents for the third climactic eruption.

While not so voluminous as the first ash flow eruptions, the third cycle ejected at least 250 cubic miles of new magma, approximately a thousand times more than the 1980 eruption of St. Helens. This immense quantity of pulverized hot rock hurled miles into the stratosphere formed a vast ash canopy that must have darkened skies over much of North America. Fallout from the Yellowstone ash column is preserved in locations as far removed as Kansas, California, and Saskatchewan. The rapid partial emptying of the magma chamber again caused its roof to collapse, forming the present Yellowstone caldera, which measures approximately thirty miles across and fifty miles long, one of the largest in the world.

The third series of ash flows probably annihilated plant and animal life over thousands of square miles of the western and midwestern United States, but the Yellowstone volcano's energies were not yet exhausted. Shortly following caldera formation, two upwelling tongues of magma arched the caldera floor to form two broad domes. Since then the volcano has erupted another 250 cubic miles of magma, equaling the quantity of that discharged during the third climactic eruption. During the last 150,000 years, extensive flows of rhyolite lava have buried much of the Yellowstone basin. The youngest lavas erupted in the park poured out only about 70,000 years ago.

Although Yellowstone's last significant eruptions happened well before the end of the Ice Age, a large reservoir of hot magma probably exists at shallow depth beneath the park. That magma may be evolving toward producing another catastrophic explosive eruption. Judging by the last three eruptive cycles, it took from about 200,000 to 600,000 years to generate the magma needed for an ash flow holocaust. The underground growth of magma was indicated at the surface by the intermittent eruption of rhyolite lava flows and domes.

It is impossible to determine whether the rhyolite eruptions of the last 150,000 years represent the end of the third caldera-forming cycle or a significant precursor of the fourth. Some evidence suggests that Yellowstone's thermal areas are growing larger, and hotter. Intense thermal activity, measurable crustal uplift of large areas of the caldera floor, and frequent earthquakes suggest that fresh magma has been injected into Yellowstone's magma reservoir since the latest eruptive period ended 70,000 years ago.

The Yellowstone volcano will erupt again. Future activity inside the caldera will probably produce rhyolitic magma, perhaps on a moderately explosive scale comparable to recent eruptions at the Mono-Inyo Craters in east-central California. Eruption of larger volumes of rhyolite may produce an extremely explosive outburst, like that which beheaded Mt. Katmai and formed the Valley of Ten Thousand Smokes in 1912. Basalt magma erupted outside the caldera is likely to form quiet effusions of lava and chains of small cinder cones. With its giant magma reservoir simmering atop a notoriously persistent hot spot, however, the Yellowstone volcano will eventually stage another caldera-forming paroxysm that could reduce large areas of western North America to a rhyolite desert.

Molten lava pouring from fountains filled Kilauea Iki lava lake in the Kilauea caldera. The flank of Mauna Loa is visible in the distance. —Donald Hyndman photo

Chapter 20

THE FIERY REALM OF MADAME PELE:
The Volcanoes of Hawaii

Not everywhere can one visit the home of a living goddess. On the Big Island of Hawaii, however, the curious can peer directly into the principal residence of Madame Pele, the Hawaiian goddess of volcanic fire.

According to ancient myth—and in many cases present belief— Pele lives in the Halemaumau fire pit, a vertically walled circular crater on the floor of the summit caldera of the Kilauea volcano, one of the most frequently active fire mountains in the world. A major feature of the Hawaii Volcanoes National Park, the Kilauea caldera, an oval-shaped collapse depression measuring two by two-and-one-half miles in diameter, is remarkably accessible even to casual

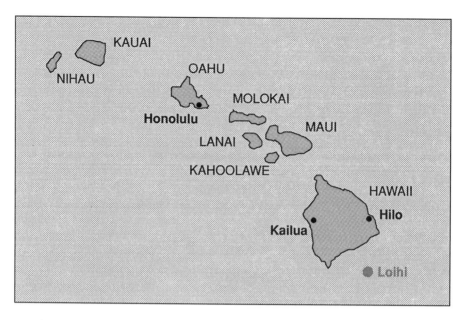

The summits of huge volcanoes rising from the Pacific Ocean floor, the Hawaiian Islands were built over a persistent hot spot in the earth's mantle. As the Pacific plate travels northwestward across the hot spot, it carries the older islands away from their magma supply, causing volcanoes on the northern islands to become extinct. At present the Big Island of Hawaii sits over the hot spot, but a vigorous new submarine volcano, the Loihi Seamount, is building directly to the south. Loihi will eventually break the ocean surface to become the newest vacation mecca in the Hawaiian archipelago.

tourists. An eleven-mile-long paved road loops around the caldera rim, dipping down to the caldera floor on the southwest side, where a short trail leads to the very edge of Pele's sulphurous abode.

The goddess, notorious for her blistering temper and unpredictable behavior, was not officially at home when I dropped by to pay my respects in the mid-1980s. She had temporarily abandoned Halemaumau to tend new fires along the east rift zone, a linear series of faults and fissures extending northeast from Kilauea's summit along the outer flanks of the volcano. Since January 1983, Madame Pele has produced floods of molten rock from several different vents along the east rift zone. As of 1990, the eruption continues unabated, the longest eruptive episode in Hawaii's recorded history. During much of this time, Pele concentrated on building a new cinder cone—Pu-u O-o (Hill of the O-o bird)—which has attained a height of over 850 feet. Although the new volcano is not easily visited, hikers can obtain an unobstructed view by following a mile-long trail to the top of Pu-u

Huluhulu, an older cinder cone near the Chain of Craters road, which has been repeatedly blocked by streams of lava flowing toward the sea.

When Pele is irritable, many practitioners of the old faith try to appease her wrath by offering gifts. Before the island population was Christianized, pigs were slaughtered as sacrifices. Today the devotee usually brings flowers, the Ohelo berries, sacred to the goddess, or even a bottle of gin—the lady is said to be particularly fond of this beverage. On a recent visit I found the Halemaumau crater rim littered with incense sticks, bouquets, including a pot of chrysanthemums, and a flask of the preferred liquor. These friendly attempts to calm Pele are not amiss, for many times during this century she has directed enormous torrents of lava over previously fertile land, converting lush forests, fields of sugar cane, and entire villages to barren deserts of black rock.

Pele is personally responsible for the remarkable fact that our fiftieth state is, literally, the fastest-growing area in the West. Between 1969 and 1974, effusions of fluid basalt constructed Mauna Ulu (Growing Mountain) on Kilauea's east slope, from which basaltic flows traveled eight miles to the seacoast, creating 200 acres of new land. Pele's current activity has similarly extended Hawaii's coastline. The entire Hawaiian archipelago, from Midway to the Big Island

Steam rises from the Halemaumau crater on Kilauea, viewed from Kilauea Overlook on the north rim of the volcano. —Donald Hyndman photo

itself, is a succession of geological monuments to Pele's creative energy. Each island is the top of a largely submerged volcano.

Hawaii, from which the whole island group takes its name, is the largest and youngest of Pele's building projects, consisting of five overlapping shield volcanoes. A volcanic shield, so called because of its supposed resemblance to a warrior's shield laid flat with the curved side up, is a broad, very gently sloping structure composed of thin, sheet-like flows of basalt lava.

The Hawaiian shield volcanoes are the largest and highest mountains on earth. The loftiest is Mauna Kea (White Mountain), which wears a snowcap from fall until late spring. Mauna Kea's visible cone

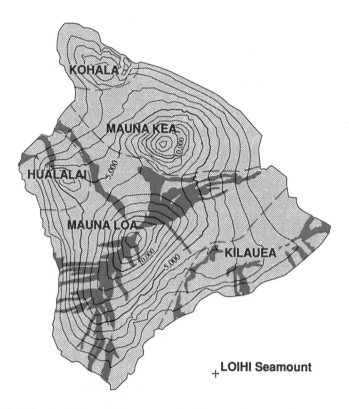

The Big Island off Hawaii is composed of five overlapping shield volcanoes that rise as high as 31,000 feet above the ocean floor. Mauna Loa and Kilauea erupt frequently, while Hualalai last erupted in 1800-1801. Red areas represent historic lava flows. The Loihi Seamount, a young submarine volcano about twenty miles south of Kilauea, will eventually break the ocean surface and form a new island.

A rift along the rim of Kilauea caldera. Halemaumau crater in the distance. —Donald Hyndman photo

stands 13,796 feet above sea level, but measured from the ocean floor it rises approximately 31,000 feet from base to summit. Only about 100 feet lower in elevation, its neighbor Mauna Loa (Long Mountain) is even larger in volume, an estimated 10,000 cubic miles, and still growing. The 1950 eruption produced an enormous outpouring of molten basalt that significantly augmented the Big Island's coastline. In 1984 Mauna Loa and Kilauea erupted simultaneously, adding still more real estate to Pele's domain and demonstrating that the goddess can operate in two places at once.

According to legend, Pele's second home is atop Mauna Loa, in a three-mile-long caldera named Mokuaweoweo. Mauna Loa's eruptions typically begin in the summit caldera and then break out on the outer slopes of the giant shield. Before the lava begins to flow, however, Pele customarily warns her people of impending danger by appearing to a few witnesses as either a beautiful young woman or an old lady in distress. The destructiveness of the forthcoming eruption is said to be determined inversely by the degree of hospitality shown to Pele during these mysterious visitations.

Following her warning appearances, Pele opens the show in Mokuaweoweo with spectacular fire fountains, towering jets of liquid rock powered by gas escaping from the molten material. These sprays of lava can reach awesome heights—one at Kilauea recently shot 1,900 feet skyward, the highest ever measured.

149

A basalt flow channel spills out of Keanakakoi crater on the south rim of the Kilauea caldera, Hawaii. —Donald Hyndman photo

A lava tree cast presents a ghostly profile at Lava Tree State Monument, southeast of Hilo, Hawaii. A basalt flow half buried and burned the tree. When the lava level dropped it left this mold of the trunk. —Donald Hyndman photo

The toe of this pahoehoe flow crept onto the Chain of Craters Road during a November 1973 eruption of Kilauea. —Donald Hyndman photo

Molten rock pouring from the fountains commonly forms a lava lake in Mokuaweoweo, which eventually drains back into underground reservoirs when this opening phase of the activity is over. During the second part of the eruptive cycle, Pele moves from the summit to the volcano's lower slopes. There she opens long fissures on Mauna Loa's flanks; from these linear fractures a chain of lava jets emerges, forming a curtain of fire, sometimes several miles in length. Later in the cycle, Pele focuses her attention on a few vents that feed rivers of molten rock that travel many miles downslope, enveloping plantations and villages in their path.

In 1984 rivers of fluid basalt threatened to engulf Hilo, the largest city on the island. As in past eruptions, the goddess relented, and the lava stopped before reaching the island's busiest port. The voluminous outpourings from Kilauea during the 1980s repeatedly blocked the Chain of Craters Road and destroyed scores of homes and other buildings on Kilauea's east flank.

It is probably only a matter of time before Hilo, as well as many other settlements and luxury resorts on the Big Island become victims of Pele's wrath. Even a cursory survey of Hawaii's topography reveals

151

that few areas have escaped Pele's ministrations. Much of the island consists of vast tracts of barren basalt, erupted so recently that they support little vegetation.

The geologic record offers no assurances that Madame Pele will not temporarily vacate her favored homes in the Halemaumau and Mokuaweoweo fire pits and reactivate old haunts or even establish new hearths at unexpected places. One largely ignored site of Pele's former habitation is the Hualalai volcano that looms above the now densely populated Kona coast. In 1800-1801 the goddess briefly revisited Hualalai to produce copious streams of lava that extended all the way to the sea. The Kona airport now sits atop one of these congealed flows. Hualalai's nearly two centuries of quiet are no guarantee that Pele will not return. When Pele again leaves Halemaumau to ignite Hualalai's fires, the Kona tourist industry, concentrated along the volcano's western foot, may be forced to relocate. Whenever she sets up housekeeping next, the goddess is certain to redesign the surrounding landscape in her favorite shade of basaltic black.

Namaka, goddess of the sea, attacks her sister Pele's lava creation on the Big Island, Pele's current home. —Donald Hyndman photo

Chapter 21

THE HAWAIIAN HOT SPOT

In Hawaiian myth the goddess Pele built her fires on other islands before settling down on the Big Island of Hawaii. Tradition states that Pele was driven from her former homes on Kauai, Oahu, and Maui by her sister Namaka, goddess of the sea.

The old myths accurately reflect geologic fact. Constructed during the last 700,000 years, the Big Island is the youngest in the Hawaiian chain and is still actively growing . Much of Mauna Loa's surface has been covered by new lava during the last few thousand years, while a full ninety percent of Kilauea's surface was formed during the last 1,000 years, some of it only yesterday. The islands become progressively older to the northwest and consequently progressively more eroded by the relentless battering of sea waves, the work of Namaka.

This pressure ridge in a 1971 Puna Coast pahoehoe flow formed as slabs of hardened basalt floating on molten lava jammed together. —Donald Hyndman photo

Competition between the two sisters represents the age-old conflict between nature's coeval powers of construction and destruction. As soon as Pele's lavas form an island above sea level, Namaka attacks it with all the fury of Pacific storms, wearing away the rock and washing it out to sea.

Traveling northwestward along the 1,500-mile-long Hawaiian archipelago to the island of Midway confirms Namaka's ultimate victory. Erosion has worn the northernmost volcanic islands below sea level, allowing coral atolls to form atop the drowned mountains. Midway, millions of years older than Hawaii, represents this late stage of the islands' evolution. Farther north a line of submarine volcanic cones, the Emperor Seamount chain, extends beneath the north Pacific until it disappears into the deep undersea trench between Kamchatka and the western tip of the Aleutian Islands.

In the sibling rivalry between Pele, the creative island-builder, and her destructive sister, Namaka has the benefit of time on her side. A restless wanderer, Pele typically finishes one construction project and then abandons it and moves on to the next. She completed Kauai almost six million years ago and Oahu about three million years later, giving Namaka ample leisure to erode these islands. Namaka has had a full seventy million years to tear down the Emperor Seamounts, which once rose as high above the Pacific Ocean surface as Hawaii or Maui stand today.

Pele's southward wanderings to her present abode in the Halemaumau fire pit offer an important clue to the geologic processes that created the entire Hawaii-Emperor chain. Like Yellowstone Plateau on the mainland, the Big Island now rises above a hot spot rooted in Earth's mantle. This subterranean concentration of heat generates unusually large volumes of basaltic magma that erupt on the ocean floor, gradually piling up to form Hawaii's island chain.

The enormous basaltic slab that forms the Pacific Ocean floor moves northwestward across the stationary hot spot at the rate of about four inches per year. As the Pacific plate glides past the hot spot, the individual volcanoes built above it are cut off from their magma supply, one after the other. Like a giant conveyer belt, the moving plate carries the volcanoes away from their underground heat source, causing their fires to sputter and die. No longer replenished by new floods of lava, the dying volcanoes are defenseless against unending erosion by storm and stream. Namaka's eroding agents cut deeply into the volcanoes' flanks and summit calderas, reducing their elevation and excavating enormous canyons like those seen today at Haleakala volcano on Maui.

The Hawaiian hot spot formed about 70 million years ago, perhaps when a large meteorite struck the Pacific Ocean basin. As the Pacific plate moves northwestward across the hot spot, the Big Island volcanoes will be carried away from their magma supply and become extinct. A new volcano, the Loihi Seamount, is already rising from the ocean floor south of Hawaii and will probably reach the surface within a few thousand years.

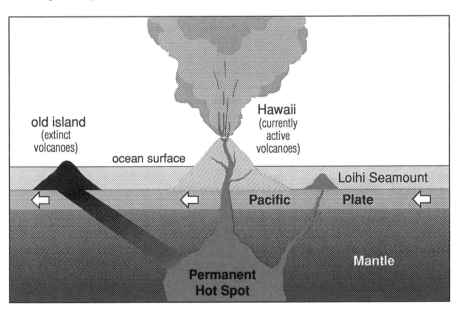

155

A frozen, crumpled pahoehoe river "flows" through brush at the northern end of the Chain of Craters Road, Kilauea caldera, Hawaii.
—Donald Hyndman photo

Hawaii's volcanic landforms represent every stage of the construction-erosion cycle. Regularly erupting immense volumes of fluid basalt, Mauna Loa and Kilauea are vigorously cone-building. Nearby Mauna Kea has entered a later stage typified by the eruption of more silicic, explosive magma that has erected clusters of overlapping cinder cones and thick lava flows at its summit, largely filling the summit caldera. Although it has been quiet for about 3,500 years, Mauna Kea is only sleeping and may awake to produce some spectacular explosive activity.

It has migrated some distance from the central hot spot, but Maui's giant Haleakala volcano is still intermittently active. Offering one of the world's most awe-inspiring vistas, Haleakala's enormous summit caldera, 10,000 feet above the Pacific, is partly filled with scores of brilliantly colored cinder cones and other youthful eruptive deposits. The dazzling hues of rust, gold, orange, and red result from the oxidation of iron-bearing minerals in the hot rock shortly after it was erupted. Although partly dissected by erosion, Haleakala erupted as recently as 1790 and is expected to erupt again.

As the older islands drift inexorably northward to their final subduction in the Aleutian Trench, Pele is already preparing a fresh

156

addition to Hawaii's volcanic family. About twenty miles directly south of Kilauea, a new volcano, the Loihi seamount, is now rising from the ocean floor. Loihi is not expected to breach the ocean surface for several thousand years, but when it does it will produce some violent steam-blast eruptions as hot magma mixes with shallow sea water. Loihi will then settle into the kind of familiar Hawaiian effusive eruptions we see today at Mauna Loa and Kilauea, providing a new volcanic paradise for our vacationing descendants.

Although the myths correctly point out that over long ages of geologic time Pele has changed her center of operations from one island to another, always traveling in a southeast direction, in reality Pele has kept a fixed abode. Her true home is the incandescent interior of the Hawaiian hot spot, an apparently inexhaustible reservoir of energy and building material.

Pele entered her permanent home about seventy million years ago, when the first of the Hawaiian-Emperor volcanic islands began to rise. As befits a goddess, she may have arrived on earth from outer space, perhaps transported here in a celestial fiery chariot. By analogy to the creation of other hot spots, Pele may have reached earth via meteorite. The meteorite impact that struck western India about

A lava tree cast at Lava Tree State Monument, near Hilo, Hawaii. The hollow core of the cast shows the texture of the burned-out tree trunk. —Donald Hyndman photo

The smooth, ropy surface of a pahoehoe lava flow. Pahoehoe flows are extremely fluid. As the lava at the surface cools and begins to solidify, the underlying lava continues to flow, rumpling the surface. —Donald Hyndman photo

sixty-five million years ago created a persistent low pressure cell in earth's mantle that generated the colossal lava flows of India's Deccan Plateau. A similar impact in southeastern Oregon about 17 million years ago may have triggered the basaltic floods and rhyolite eruptions that form the Columbia River Plateau and the Snake River Plain-Yellowstone system.

If Pele indeed traveled from somewhere in the solar system to smash into the mid-Pacific, the results would have been catastrophic almost beyond imagining. A stony meteorite about six miles in diameter traveling at ten to twenty miles per second would have 10,000 times the power of all the world's nuclear weapons. Within seconds of impact, a gigantic fireball thousands of miles in diameter would soar many miles into the stratosphere. Superheated winds would rake the planet for hours, searing the world's vegetation and igniting forest fires on almost every land surface.

Heat at the target site would vaporize as much as five trillion tons of ocean water, creating a pillar of vaporized water fifteen or twenty miles across that would soar many miles above a boiling sea. Enormous

158

sea waves would rush outward from the impact area, crashing into coastlines all around the Pacific basin and scouring them clean of plant and animal life. An enormous canopy of dust would envelop the globe, shutting out the sun. Soot from global fires would compound the gloom.

The column of water vapor permeating the atmosphere would soon condense as rain or snow, initiating a total precipitation of about 3,500 feet. Rainfall of biblical proportions, averaging 200 to 300 inches per day, would drown the incinerated earth.

While the cloud cover and dust pall would lower global temperatures for several months, the long-term effects of the impact may have created a greenhouse effect, heating up the planet. Earth-wide extinction of species would have been hastened by hot nitric acid that rained from the sky, dissolving the shells of marine organisms.

Vast tides of molten rock erupting from the impact site on the ocean floor would affect ocean currents, temperatures, and world climate for long millennia. The appearance of new volcanic islands where none had existed before, the first of the Hawaii-Emperor archipelago, would be among the least of the impact's life-destroying consequences.

Like the rugged beauty of Yellowstone National Park, Hawaii's tropical paradise is the relatively benign product of cosmic violence. Todays eruptions of glowing lava are but faint echoes of the chaotic event that may have accompanied Pele's arrival on earth.

Mount Redoubt, elevation 10,197 feet, towers above the western shore of Cook Inlet. Like Spurr, Iliamna, and Augustine, Redoubt is one of several explosive volcanoes whose ashfalls have darkened skies over Anchorage and interfered with air and sea travel in southeastern Alaska. Redoubt began its latest series of eruptions in December 1989.

—U.S. Geological Survey, Austin Post photo

Chapter 22

ALASKA'S MOUNTAINS
OF ICE AND FIRE

Famous for its vast icefields, tidewater glaciers, and bitter winter cold, Alaska is also a state of volcanic fire. Second only to Indonesia as the most volcanically active region on earth, Alaska's numerous volcanoes are among the world's most violently explosive.

Every year Alaskan fire-mountains produce an average of one to three eruptions, some of which have far-reaching effects. The largest eruption of the twentieth century destroyed the summit of Mount Katmai in 1912 and created the spectacular Valley of Ten Thousand Smokes, now the central feature of Katmai National Park (Chapter 23). That eruption shot six cubic miles of ash high into the air, generating an atmospheric pall that for a few years significantly lowered temperatures in the northern hemisphere.

In December 1989 Mt. Redoubt, a 10,197-foot volcano in the Aleutian Range 110 miles southwest of Anchorage, began a series of vigorously explosive eruptions that blew ash six miles above its glaciated crest. Caught in the huge ash cloud, a Boeing 747 passenger jet experienced failure of all four engines, causing it to plunge 13,000 feet before regaining power and making an emergency landing. No one was killed in the incident, but it illustrates the danger that Alaska's ash-producing volcanoes pose in a state that relies almost entirely on air traffic for transportation and supplies. The U. S. Postal Service temporarily suspended all mail deliveries to and from Alaska because of the hazard from ash clouds, leaving Christmas mail and groceries for northern, central, and western parts of the state undelivered.

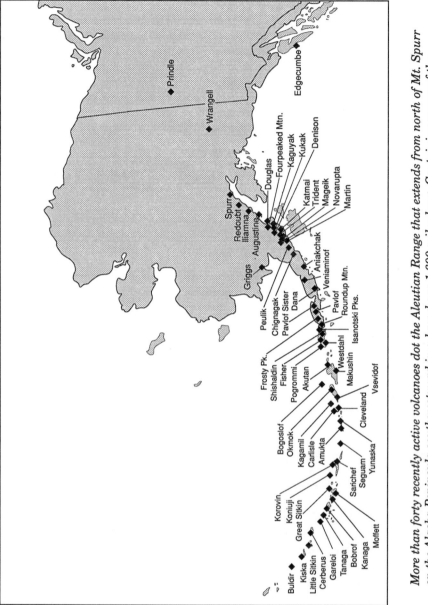

More than forty recently active volcanoes dot the Aleutian Range that extends from north of Mt. Spurr on the Alaska Peninsula southwestward in a broad arc 1,600 miles long. Containing many of the world's most frequently and violently explosive volcanoes, the Aleutian chain is created by the subduction of the northern Pacific Plate beneath northwestern North America.

Viewed from the south, Mount Redoubt steams vigorously after a recent eruption. Explosive outbursts in 1966-1968 and 1990 caused extensive melting of the volcano's icefields and sent voluminous floods and mudflows traveling down its flanks.
—U.S. Geological Survey, Alaska Volcano Observatory photo

Early in 1990 explosions shattered Redoubt's summit dome, sending avalanches of hot rock down the volcano's icy slopes and triggering floods and mudflows that swept down the Drift River valley toward Cook Inlet. Five times the volume of the largest mudflow produced by St. Helens in 1980, the surging mass of water-saturated rock debris partly overflowed a protective levee and threatened the Drift River oil terminal. Located twenty miles downvalley from Redoubt, the oil storage facility normally processes 30,000 barrels a day, making it Alaska's second largest source of oil. Citing the danger to oil tanker traffic in Cook Inlet, the U.S. Coast Guard ordered the terminal closed.

Redoubt's eruptions repeatedly sent ash clouds drifting over Anchorage and other Alaskan cities, including Sitka, 650 miles southeast of the volcano. Redoubt was previously active in 1966-1968 when mudflows also inundated the Drift River valley, including the site where the 37-million-gallon storage facility was later built.

In 1986 the Augustine volcano, located on a tiny island in Cook Inlet about 180 miles southwest of Anchorage, spewed ash over a wide area, darkening skies over Alaska's largest city. Besides interfering with

View of Rust Slough and portions of the Drift River Oil Facility. Mudflows generated by Redoubt's 1990 eruptions inundated the slough channel, causing floods to overflow the facility tank containment dikes. —U.S. Geological Survey Alaska Volcano Observatory photo

shipping and airplane flights, Augustine's pyrotechnics are potentially dangerous because they can generate massive sea waves that can swamp ships and damage buildings in low-lying coastal areas. The eruption of 1883 produced waves up to thirty feet high that struck the fishing village of English Bay about fifty miles east on the Kenai Peninsula. Since 1963, Augustine has erupted with increasing frequency and may eventually prove a serious threat to Homer and other towns along or near Cook Inlet.

About forty-seven of Alaska's eighty-plus volcanoes have erupted since the mid to late 1700s, when Russian and other European explorers first reported seeing ice-clad peaks blaze and smoke. Pavlof, rising to an elevation of 8,900 feet about forty miles northeast of Cold Bay, is the most active volcano in North America. The highest of a trio of young composite volcanoes that includes Little Pavlof and Pavlof Sister, it has staged nearly thirty eruptions during historic time, the latest in 1988.

Shishaldin, which towers more than 9,900 feet above the northeast corner of Unimak Island, directly southwest of the Pavlof group, is almost as frequently active. One of the world's most exquisitely symmetrical mountains, Shishaldin's snowy cone rivals that of Japan's Fujiyama or the Philippines' Mount Mayon. Between eruptions, Shishaldin commonly trails a billowing plume of steam from its icy summit.

All of these fire-mountains are part of an unbroken volcanic chain that begins north of Mount Spurr on the Alaska Peninsula and extends the entire length of the Aleutian Islands—a distance of 1,600 miles! The northernmost link in the circum-Pacific Ring of Fire, the Aleutian Range volcanoes form a remarkably long and narrow arc that ultimately merges on the west with the Kamchatka volcanic belt in the Soviet Union.

The Aleutian Range is a chain of volcanoes that lies parallel to the line of collision of two plates of the earth's lithosphere. Moving inexorably northwestward, the Pacific plate thrusts against the lighter crustal slab carrying the Bering Sea on its back and sinks beneath it. The descending, water-saturated slab of basaltic ocean floor heats as it sinks until the water boils out of its rocks as superheated steam. The

A smoldering island "lighthouse" in Alaska's Cook Inlet, St. Augustine volcano has erupted with increasing frequency during recent decades. The 1883 outburst generated destructive tsunamis along the Kenai Peninsula, fifty miles to the east, while the explosive eruptions of 1963-64, 1976, and 1986 blanketed Homer, Anchorage, Valdez, and other parts of southcentral Alaska with gritty ash. —U.S. Geological Survey, Austin Post photo

addition of water lowers the melting temperature of the mantle rock, which partly melts to form basaltic magma that rises into the overlying crust. Where the continental crust is old and granitic, containing a high proportion of quartz and feldspars, the rising molten basalt heats it to produce granitic magma. If the granitic melt has a high percentage of water, it rises to erupt explosively at the surface.

The Alaskan volcanoes are particularly explosive, and therefore dangerous, because of the granitic composition of their magma. The mid-oceanic volcanoes, such as those in Hawaii, erupt basalt almost exclusively. Because basalt has a low silica content, the gases dissolved in the magma can escape easily when erupted, forming the brilliant red and orange sprays of lava that characterize a Hawaiian eruption. By contrast, much of the magma produced near a subduction zone has a much higher silica content, forming andesite, dacite, or even rhyolite. Hot gases entrained in a silicic magma do not escape until they reach the surface, when the confining pressure of overlying rock is removed.

When the water dissolved in magma escapes to form steam, the most abundant volcanic gas, its volume expands and increases a thousandfold. The sudden and rapid gas expansion, which blows magma into millions of fragments, causes the tremendous explosions typical of Alaska's volcanoes. Instead of flowing out quietly in liquid streams of lava, as happens in Hawaii, the magma jets high into the stratosphere, spreading out in a huge canopy of hot ash that shuts out the sun and turns day into night.

When such volcanoes erupt silica-rich lava such as dacite or rhyolite, they commonly also produce pyroclastic flows, the most deadly of all volcanic phenomena. Traveling up to 100 miles per hour, pyroclastic flows typically follow topographical depressions, such as stream valleys, but the hot ash clouds accompanying them cover much broader areas, sweeping over ridgetops and devastating adjoining areas.

When red-hot silicic lava erupts in thick, sticky masses high on a mountain, it typically collapses and avalanches downslope to create lithic pyroclastic flows. In 1883 this kind of eruption at the Augustine volcano triggered pyroclastic flows that plunged into the sea, creating the tsunamis that swept Cook Inlet. Geologists regard a future repetition of this event as one of the major geologic hazards to some of Alaska's principal shipping lanes.

At least nineteen times during the last few thousand years Alaskan volcanoes erupted so violently that their former summits collapsed,

A land of ice and fire: the many volcanoes of Alaska's eastern Aleutian Range tower above a rugged glaciated mountainscape. A composite cone built of lava flows and fragmental rock, Mount Iliamna, elevation 10,016 feet, has produced at least six eruptions during historic time, the last in 1947. Iliamna's near neighbor, Mount Spurr, erupted violently in 1953, shrouding Anchorage and much of southeast Alaska under fine gray ash.
—U.S. Geological Survey, Austin Post photo

forming large calderas. One of the largest is Aniakchak Crater, the centerpiece of a newly created national monument located at the west end of the Alaska Peninsula about 400 miles southwest of Anchorage. Accessible only by pontoon planes that can land on the lake occupying part of the caldera floor, Aniakchak, six miles in diameter, is surrounded by cliffs up to 2,000 feet high.

After Aniakchak's collapse about 3,400 years ago, activity inside the caldera produced lava flows, domes, and a massive cinder cone, Vent Mountain, which erupted as recently as 1931. Aniakchak River drains the caldera lake through a breach in the eastern rim, where it plunges downslope in a churning torrent of white water that merits its recent official designation as a National Wild River.

The Okmok volcano, on Umnak Island, is topped by two overlapping summit calderas, each about six miles in diameter. The first summit collapse happened about 8,250 years ago, the second less than 2,400 years ago. Both eruptions ejected silicic magma as huge ash columns and voluminous pyroclastic flows that devastated areas many tens of miles from the volcano.

Milder explosive eruptions continue intermittently. In 1817 the Okmok volcano hurled massive lava blocks at least forty-five miles and blanketed parts of Unalaska with fifteen centimeters of ash. During World War II, earthquakes, ashfall, and reports of glowing lava forced the evacuation of a U. S. Army base on the east side of Umnak Island. Sporadic ash eruptions took place throughout the 1980s.

One of the most powerful caldera eruptions since the end of the Ice Age occurred at Fisher caldera near the west end of Unimak Island. About 9,100 years ago the volcano blew out immense quantities of dacite pumice, forming pyroclastic flows mobile enough to travel miles across lowlands and climb nearly 1,500 feet to sweep over distant ridgetops. The resulting collapse depression measures approximately eight by fourteen miles. A small eruption may have occurred as recently as 1826-1827.

More frequently active, the Veniaminof caldera on the Alaska Peninsula formed about 3,700 years ago. An enormous composite cone 100 cubic miles in volume, the volcano erupts both pyroclastic material and fluid streams of lava. In 1892 it blanketed southwest Alaska under an extensive ashfall and remains sporadically active. During the 1980s activity in Veniaminof caldera included vigorous lava fountaining, lava flows, the building of a new cinder cone, and the ejection of ash plumes several miles into the air.

Mount Wrangell

Southeast of the main Aleutian arc, the Wrangell Mountains, the jewels of central Alaska, are a compact cluster of volcanic peaks that include Mounts Drum, Sanford, Wrangell, and Jarvis. Among the largest volcanic cones in the world, they probably grew rapidly through voluminous outpourings of lava during the last half million or so years. Rising more than 14,000 feet above sea level, Mount Wrangell is an enormous shield volcano whose broad summit is indented by a caldera about 2.5 by four miles in diameter. An older, larger caldera perhaps twelve miles across extends northeast from the present summit depression. Three smaller craters formed along the younger caldera rim.

Like those of Rainier and Baker in Washington state, Wrangell's ice-filled craters contain long caves or passageways created as volcanic heat partially melts the crater icepack. A significant increase of heat and steam emission after 1965 radically transformed the ice cave system. By 1986 about eighty-five percent of the pre-1965 ice in North Crater had melted. Steam issuing from crater fumaroles occasionally forms plumes several thousand feet high.

The sudden rise in Wrangell's heat production may relate to the 1964 Prince William Sound Earthquake. An earlier magnitude 8-plus quake, that in Yakutat Bay in 1899, triggered a similar period of vigorous steaming. Earth scientists suggest that some of Alaska's volcanoes are directly affected by plate movement. Pavlof and Veniaminof, for example, seem particularly sensitive to movement along the Aleutian subduction zone. Increased activity at these volcanic centers may precede or follow large earthquakes.

Each Alaskan caldera records a violently explosive eruption that expelled many cubic miles of molten rock and devastated hundreds to many thousands of square miles. Twelve calderas, from Mount Wrangell on the Alaskan mainland to Little Sitkin in the western Aleutian Islands, are classified as geologically "restless," the source of historic eruptions or potentially centers of renewed explosive activity. Several, including Okmok and Wrangell, produced at least two or more violent episodes of caldera formation, suggesting that some may again erupt catastrophically.

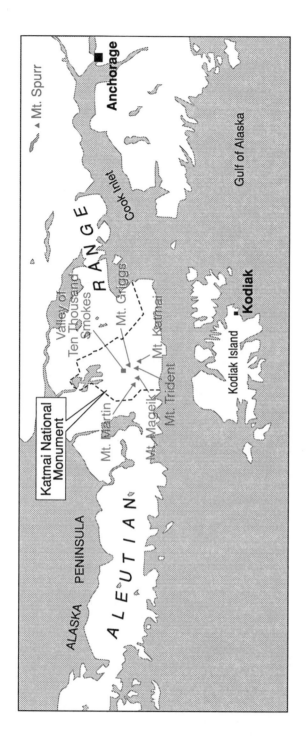

Katmai National Park on the Alaskan Peninsula was the site of this century's largest and most violently explosive volcanic eruption. In 1912 a new vent, Novarupta, at the foot of Mount Katmai ejected more than eight cubic miles of incandescent ash, much of which was propelled high into the stratosphere. Ash suspended in the stratosphere reduced the amount of sunlight reaching the northern hemisphere, resulting in a pronounced drop in average temperatures for several years. Katmai is but one in a cluster of active volcanoes that forms part of the Aleutian Range.

Chapter 23

THE VALLEY OF TEN THOUSAND SMOKES:
Katmai National Park

"But where are the smokes?"

The red-haired lady from Australia had a point. Our group stood on a promontory overlooking Alaska's desert-like Valley of Ten Thousand Smokes, created in 1912 by one of the greatest volcanic eruptions in recorded history and now the central feature of Katmai National Park. When discovered in 1916, the valley contained thousands of miniature volcanoes spewing columns of steam 500 to 1,000 feet into the air. This profusion of roaring steam jets inspired their discoverer, Robert Griggs, leader of the several National Geographic Society expeditions to the Katmai region, to name the area for its myriad fumaroles. To the lady's disappointment, seventy-five years after their creation, scarcely a smoke remained.

Early investigators like Griggs thought that the valley fumaroles tapped an underground reservoir of molten rock and would outlast Yellowstone's geysers and hot springs. By 1930, fumarole activity had drastically declined. Today, geologic detective work has solved the mystery of the vanishing smokes and in the process also revealed the cataclysmic series of events that led to their creation.

Located on the thinly populated Alaska Peninsula about 290 airline miles southwest of Anchorage, Katmai National Park at 2,800,000 acres is one of the nation's largest legally protected wilderness areas. Its southeastern boundary is Shelikof Strait, which separates Kodiak Island from the mainland. A rugged terrain of glacier-carved lakes, white water streams, conifer-timbered lowlands, and ice-shrouded peaks, Katmai National Park encompasses a

Created in 1912 by the collapse of Mount Katmai's former summit, the Katmai caldera is about three miles in diameter and measures 3,700 feet from the caldera floor to the highest point on the rim. The small glacier (right center) began to form inside the caldera wall shortly after the 1912 eruption. The caldera lake, rising at an average rate of sixteen feet per year, is expected to overflow the east caldera wall about the year 2050.
—U.S. Geological Survey, Austin Post photo

picturesque segment of the Aleutian Range, including Mount Katmai and at least a dozen other volcanoes.

Alaska's volcanoes are typically explosive, but the 1912 catastrophe, in size, power, and volume of material ejected, dwarfs any other eruption in the world during this century. For two and a half days, between the afternoon of June 6 and the early morning of June 9, a new vent at the southwest foot of Mount Katmai disgorged between eight and nine cubic miles of pyroclastic rock. This compares to the 0.3 cubic mile of ejecta from Mount St. Helens during all its 1980 activity.

About five cubic miles of tephra, airborne ash or rock fragments, were hurled into the atmosphere and stratosphere, as enormous mushroom clouds of pumiceous ash turned day into night over much of southern Alaska. Kodiak village on Kodiak Island, 100 miles to the southeast, was blanketed under a foot of tephra, while ash fell visibly over Washington state's Puget Sound, 1,500 miles southeast of the volcano. Residents of Juneau, 750 miles distant, repeatedly heard the thunder of colossal explosions.

Following the initial vertical ejection of tephra, a massive ash flow swept outward from the erupting vent, the largest erupted on land during historic time. The main arm of the ash flow poured eleven miles downvalley, was deflected by a glacial moraine and continued another three and a half miles, charring and burying timber. Clusters of trees growing near the vent were engulfed and carbonized. Temperatures in the incandescent mass,approximately four cubic miles in volume, were high enough, up to 950 degrees centigrade, to weld the hot rock fragments. Although most pyroclastic flows travel rapidly downslope, commonly up to 100 miles per hour, the 1912 ash flow was relatively sluggish, at least near its terminus. One geologist suggested that a healthy wolf could have outrun it—if the blast of hot air preceding the pyroclastic flow had not incinerated it first.

Looking across the steaming crater of Mount Mageik toward the Valley of Ten Thousand Smokes, Katmai National Park. During the 1912 eruptions of Novarupta, a new crater at the foot of Mount Katmai, a huge pyroclastic flow of hot rhyolitic ash buried the valley floor. Buried streams and groundwater flashed into steam, which, combined with gases escaping from the hot ash, produced the myriad steam plumes that gave the valley its name. As the ash deposits gradually cooled, most of the "smokes" disappeared. —U.S. Geological Survey, Austin Post photo

The voluminous outpouring of hot rock fragments, ranging in size from dust grains to huge pumice bombs tens of feet in diameter, into the sparsely forested valley west of Katmai produced the Valley of Ten Thousand Smokes. Before the eruption, the valley had contained marshes or swampy areas, as well as several glacier-fed streams. When inundated by the pyroclastic flow, stream and groundwater were transformed into steam, triggering secondary explosions under the pyroclastic deposit and blasting open craters in the flow surface, particularly over buried creek beds. More commonly gas escaped through crevices and fissures in the pyroclastic flow, creating thousands of new vents throughout the ash-filled valley. Although steam was most common, other volcanic gases hissed and sputtered from countless fumaroles, veiling the region in a sulphurous mist appropriate to Dante's Underworld. Except for those centered near the eruptive source, now plugged by the jagged lava dome which Griggs christened Novarupta, the valley fumaroles were rootless. Despite Griggs' belief, they did not originate in underlying magma but from the hot volcanic deposits in contact with creek and groundwater. As the pyroclastic flow material degassed and cooled, the smokes gradually disappeared.

Turbulent streams, temporarily displaced by the thick ash flow, have since cut steep-walled ravines through the deposits, revealing cross-sections of its internal structure. Today's visitors can hike through the typically rain or windswept valley, usually amid abrasive clouds of dust kicked up by the gusts, and observe the interior of 1912 fumaroles exposed by stream and wind erosion. Commonly stained vivid hues of rust, orange, gold, or white, these "fossil fumaroles" testify to the abundance of heat and gas that for decades seethed from the valley fill.

Katmai's truncated summit now holds a brilliant blue-green lake surrounded by multicolored lava cliffs that rise up to 3,700 feet above the lake floor. Because of its newly formed caldera, three miles in diameter, Griggs and other early researchers thought that Mount Katmai was responsible for the 1912 outburst. Geologists later concluded that Katmai's summit collapse was incidental to the main event. Because of the complete absence of eyewitnesses near the eruption site and the extremely violent nature of the explosive activity, the precise eruptive sequence has been difficult to determine. Wes Hildreth, of the U. S. Geological Survey, has studied the valley deposits and recently produced a comprehensive interpretation of the valley's formation.

The eruption began with violent ejection of rhyolitic magma that rose in towering columns of white-hot ash from fissures about six and a half miles northwest of Katmai's summit. As the eruption progressed, the chemical composition of the magma changed from rhyolite to less silica-rich dacite, and finally to andesite. This progression from white or pink rhyolitic pumice to dark andesite fragments reflects a progressive decrease in the magma's silica content and a corresponding increase in iron and calcium. It appears that the new vent at Novarupta, which initially produced rhyolitic pumice, eventually tapped a reservoir of liquid andesite beneath nearby Mount Trident. Hildreth infers that magma from Trident moved to fill a posteruption void beneath Novarupta, in turn permitting magma from Katmai to move toward Trident. When Katmai's magma supply was thus drained away, internal support was withdrawn from Katmai's former summit, which collapsed to form the present caldera.

Kaguyak caldera, Katmai National Park, looking toward Shelikof Strait. A small-scale version of Alaska's many volcanic collapse basins, Kaguyak stands 2,800 feet high and contains a lake about 1.8 miles in diameter. Following the explosive eruptions that triggered the collapse of Kaguyak's former summit, quieter eruptions of viscous lava erected steep lava domes along the caldera's inner rim (right center) and formed the tiny island near the lake center. Located about sixty miles east of Brooks Lodge, Katmai Park's single visitor facility, Kaguyak can be visited only by float plane.
—U.S. Geological Survey, Austin Post photo

Some of the earthquakes that jolted west-central Alaska during the eruptions were probably triggered by Katmai's collapse and subsidence of the depression surrounding Novarupta. After the last ash flows and dacite airfall pumice erupted, a stiff, pasty mass of gas-poor lava clogged the principal vent, creating the Novarupta dome.

If they are willing to look, visitors to the Valley of Ten Thousand Smokes can still find active fumaroles. Besides those associated with the Novarupta caldera, conspicuous steam jets issue from Mount Griggs and from three other composite volcanoes that form a semicircle above the valley head. Glacier-draped Mount Trident erupted tephra and several thick lava flows between 1953 and 1968 and is still steaming vigorously. Good weather flights over the craters of nearby Mounts Martin and Mageik are rewarded by close-up views of their bilious, yellow-green crater lakes surrounded by pulsating steam vents. After a spectacular air tour of Katmai National Park, even the Australian lady was convinced that it offers a wide variety of unspoiled scenery—still being shaped by volcanic fire and glacial ice—that more than compensates for the lost smokes.

Chapter 24

VOLCANIC HAZARDS
IN THE WEST:
An Overview

Since the Ice Age ended about 10,000 years ago, large volcanic eruptions have occurred in at least eleven western states. Besides Hawaii and Alaska, which contain some of the world's most frequently active volcanoes, many locations in the Pacific Northwest, the Rocky Mountain interior, and the southwest have a high potential for future activity. California alone has at least seventy-six volcanoes that have erupted, some repeatedly, during the last 10,000 years.

Most future eruptions will resemble those of the recent geologic past and will probably take place at or near the sites of the most recent activity. The kinds of eruptions that will occur and the potential danger to human life and property they represent largely depend on the volcano's proximity to population centers and the type and volume of the magma it erupts.

The least threatening activity will consist of small eruptions of basalt lava in sparsely inhabited areas of the Basin and Range province or adjacent parts of the southwest. As in the past, most of these isolated events in eastern California, Nevada, Utah, Arizona or New Mexico result from the stretching, thinning, and fracturing of the earth's crust as North America continues to move westward. Small quantities of basalt magma leak to the surface through these extensional fractures in the crust and typically build cinder cones composed of fragmental rock. One of the most common landforms from Arizona and New Mexico to Idaho and Washington, they usually line up along fracture zones like rows of oversized ant hills.

Sites of volcanic eruptions in eleven western states during the last 5 million years. Notice that well-defined volcanic zones extend southeast from the Cascade Range through eastern California and western Nevada. Another zone of intense activity extends from eastern Oregon across Idaho's Snake River Plain into northwestern Wyoming, site of the Yellowstone volcano. Dark red marks dacite or rhyolite eruptions, sites of potentially violent explosive activity.

Relatively young cinder cones are so numerous and so widely distributed throughout the West that new ones are certain to appear in the relatively near future. Besides those in the Cascade Range of Washington, Oregon, and northern California, the youngest cinder cones on the United States mainland are Arizona's Sunset Crater and the Mono-Inyo chain in east-central California. More than a dozen of California's cinder cones erupted during the last 600 years.

When a new cinder cone is born, it will probably behave like Paricutin, the volcano that suddenly materialized in a Mexican cornfield in 1943. Within a week the new volcano grew 500 feet in height. After a year of almost continuous activity, it rose over 1,000 feet above its base. Before it abruptly ceased erupting in 1952, Paricutin stood 1,300 feet high and had poured out massive lava flows from vents at its foot that buried two nearby villages. Although no one was killed, more than 4,500 people lost their homes and farms.

Paricutin was the first North American volcano whose entire life cycle, from birth to apparent death, was studied by an international team of earth scientists. Although a Mexican phenomenon, the young fire-mountain revealed much about the geologic processes that created similar landforms throughout the western United States. Even when only moderately explosive, cinder cones may discharge heavy ashfalls that can adversely affect a region many miles downwind, as Paricutin did when its ash plume blanketed Mexico City, almost 200 miles distant.

Sunset Crater in Arizona, the youngest cone in an older volcanic field, also shrouded many square miles in ash, greatly enriching the soil later cultivated by local Indians. Even a small cinder cone

Sites of volcanic eruptions in the western United States during the last 10,000 years. The most frequently and explosively active volcanoes are in the Cascade Range of Washington, Oregon, and northern California and the Mono Lake-Long Valley area of east-central California. Other young, potentially active volcanic centers include Idaho's Snake River Plain, Arizona's San Francisco Mountains, the Western Grand Canyon and Sunset Crater, and scattered volcanic fields in Utah, New Mexico, Colorado and southeastern California. —Modified from Explosive Volcanism, National Academy Press, 1984

bursting into life near a population center could destroy or damage hundreds of buildings, highways, and other structures. The formation of another volcano near Pilot Butte, a prominent cinder cone at Bend, central Oregon's largest city, could cost Oregonians tens of millions of dollars. A much younger cinder cone, Lava Butte, stands beside Highway 97 only ten miles south of Bend. Marking the northern end of a long fault zone that has repeatedly erupted voluminous lava flows during the last 6,000 years, Lava Butte may be joined by new volcanic siblings at any time.

Very large volume eruptions of basaltic lava, such as the long flows from the Great Rift zone on the Snake River Plain, may occur at several different volcanic centers in Idaho, northeastern California, central Oregon, and perhaps northern or eastern Arizona or northern New Mexico. The broad flanks of Medicine Lake volcano or the adjacent Modoc Plateau in northeastern California are particularly likely to produce flows of basalt lava.

Except near their source, most lava flows do not move fast enough to endanger human life. Far more hazardous are the explosive eruptions that characterize subduction zone volcanoes. Fueled by the off-shore subduction of the Juan de Fuca Plate, the Cascade Range volcanoes have produced numerous cataclysmic eruptions during the last several thousand years. In calculating potential volcanic hazards, earth scientists assume that the largest probable event will be a repeat of the eruption that beheaded Mount Mazama about 6,900 years ago. That eruption ejected approximately forty cubic miles of pumice and blanketed more than 500,000 square miles with orange-beige ash. Depending upon wind direction at the time, settlements within a radius of about sixty miles of the volcano could be buried or severely damaged. Judged by its recent past, Washington's Glacier Peak is also capable of erupting catastrophically. Ashfalls from previous eruptions cover large areas of the Pacific Northwest, while mudflows extend many tens of miles downvalley into areas which are now inhabited.

The most frequently explosive Cascade volcano, St. Helens, has repeatedly ejected immense volumes of dacite pumice, depositing vast ash blankets extending northeastward across Washington, Idaho, Montana, British Columbia, and even to Alberta. Erratic winds carried ashfalls from other eruptions southward over Oregon and parts of Nevada. St. Helens' current eruptive cycle may culminate in another violent expulsion of superheated gas and molten rock, propelling more pyroclastic flows into Spirit Lake and down several river valleys heading on the mountain, thus triggering additional mudflows that will inundate communities downstream.

VOLCANOES AND WORLD CLIMATE

Explosive eruptions that inject large volumes of ash into the atmosphere can severely affect global climate. Volcanic dust acts as a screen that prevents the sun's heat from reaching earth's surface but also allows it to escape easily. By blocking solar radiation in different amounts at different places, a dust veil could create significant temperature differences between the tropics and higher latitudes, generating turbulent winds and storms. Dust particles can also function as nuclei to form ice crystals in the upper atmosphere, increasing global cloudiness and further lowering temperatures.

Eruptions in Iceland and Japan in 1783 significantly chilled the northern hemisphere, but the eruption of an Indonesian volcano in 1815 had even more adverse results. Tambora volcano ejected so much ash into the stratosphere that total darkness prevailed for three days hundreds of miles from the volcano. The eruption killed 10,000 people and another 82,000 died of starvation and disease. Millions more in Europe and North America suffered from the cooling effects of Tambora's deadly ashcloud. Snow blanketed New England in June and frosts blighted crops every month of the growing season. Heavy rains fell in Ireland, England, and northern Europe from May through October, marking 1816 as "the year without a summer."

In 1883 another Indonesian volcano, Krakatau, produced a huge ash eruption that also lowered global temperatures. A study of average global temperatures since the 1880s reveals that after a catastrophic volcanic eruption temperatures typically drop for a period of three to five years, the time it takes for dust to clear from the atmosphere. The eruptions of Katmai-Novarupta in 1912, Bezymianny in 1956, and possibly El Chichón in 1982 also depressed temperatures in northern latitudes.

The quantity of ash necessary to reduce solar radiation by twenty percent is remarkably small, much less than that erupted by St. Helens in 1980. One estimate suggests that if only a few hundred million tons of ash were ejected high enough into the stratosphere every two years, it could consistently lower the world's average mean temperature by about ten degrees. This reduction would significantly extend earth's permanent snow fields and perhaps stimulate expansion of glaciers. Some earth scientists believe that the Ice Age was initiated by repeated large-volume ash eruptions over a long period of time.

The 1985 eruption of Nevado del Ruiz volcano in Columbia was much smaller than that of St. Helens in 1980, but it claimed the lives of nearly 25,000 persons. A comparatively small quantity of hot rock ejected onto the volcano's summit icecap caused meltwater floods to race downslope, picking up large volumes of soil and loose rock debris as they plunged into adjacent valleys. The mixture of hot ash, fragments of old rock, and snowmelt formed a deadly wall of mud that swept over the entire townsite of Armero, entombing most of its 22,500 citizens. Smaller Columbian towns, along with many of their inhabitants, were similarly buried.

Today tens of thousands of persons live directly downvalley from Cascade volcanoes that, like Nevado del Ruiz, are high, steep, and crowned with thick icefields containing millions of gallons of water. During its 1980 eruptions, huge mudflows traveled down almost every stream valley heading on St. Helens, destroying logging camps, bridges, roads, and about 200 houses and other structures. In recent centuries, similarly destructive mudflows have swept many miles down valleys near Rainier, Baker, Glacier Peak, Hood, Shasta, Lassen, and other Cascade volcanoes. As Columbia's Nevado del Ruiz revealed, even small eruptions can generate mudflows that kill many thousands of persons.

Whereas basalt lava tends to flow out quietly, magmas with a higher silica content tend to retain dissolved gases until they reach the surface. Most of North America's extremely violent eruptions involve magmas composed of more than 64 percent silica by weight. At present it seems that large reservoirs of high-silica lavas, such as dacite or rhyolite, are evolving beneath several Cascade volcanoes, suggesting that one or more of them may eventually produce a cataclysmic eruption similar to that which destroyed ancient Mazama and formed Crater Lake.

Highly silicic and possibly gas-rich magma has developed beneath the western section of Lassen Park, Shasta, and the central Medicine Lake volcano, all in northern California. In central Oregon, the South Sister and Newberry volcanoes have both erupted high silica lavas during the last 2,000 years, evidence that a potentially explosive magma may now be simmering underground. Glacier Peak, composed of dacite lava, may also be preparing a caldera-forming event and Mazama, Crater Lake, may eventually stage another cataclysmic outburst.

Alaska's forty-plus historically active volcanoes are vigorously explosive, typically expelling towering clouds of ash that cover thousands of square miles. Anchorage, with a population exceeding 250,000, has had its skies polluted by ashfalls half a dozen times during the last few decades. At least nineteen times during the last 10,000 years, Alaskan volcanoes have produced eruptions comparable to the one that created Crater Lake, the latest in 1912.

Augustine volcano in lower Cook Inlet has increased the rate of its eruptions during recent years and may eventually produce a giant blast like that of Krakatau in 1883. An island volcano in Indonesia, Krakatau ejected so much pumice that its summit collapsed, generating huge sea waves that ravaged the coasts of Java and Sumatra and drowned 36,000 persons. Augustine or one of the many Aleutian Island volcanoes could explode with similar force and send great waves smashing into low-lying coastal areas hundreds of miles away.

Eruptions in Hawaii are normally quiet outpourings of basaltic lava, first at the volcano's summit caldera and later along the lower slopes. Lava flows, which are commonly fluid and voluminous, constitute the major threat to property. Three times during this century the chief port of Hilo has been threatened by long flows from Mauna Loa, most recently in 1984. A high sustained rate of lava eruption may eventually create flows that will crush, burn, and bury much of the city. Kilauea's current eruptive cycle, which began in January 1983, is the largest Hawaiian eruption in history. Pouring out lava at an average daily rate of half a million cubic yards, by mid-1990 Kilauea had destroyed about 130 houses, plantations, forested areas, and miles of roadway, mostly in or near the Kapaahu Homesteads and Royal Gardens developments on the volcano's east flank. Comparable eruptions nearer a thickly settled area would exact immeasurably higher property losses.

Although most Hawaiian volcanoes erupt quietly, they can also explode without warning. In 1924 ground water penetrated Kilauea's internal plumbing beneath the Halemaumau lava lake, triggering a succession of steam-blast explosions that hurled large blocks of lava thousands of feet from the crater. No one was killed then, but in 1790 many Hawaiian warriors died when a pyroclastic surge rolled outward from Kilauea's summit and engulfed an army. About 2,000 years earlier a series of explosive summit eruptions produced similar pyroclastic surges that swept at least six miles downslope, demonstrating that this "drive-in" volcano may again erupt dangerously. Mauna Kea, Haleakala, and dormant vents near southeastern Oahu, Hawaii's most densely populated area, are also likely to erupt explosively.

Although most catastrophic eruptions have occurred near subduction zones in Alaska and the Pacific Northwest, the very largest have taken place in the continent's interior far from plate boundaries. The Yellowstone volcano, generated by a persistent hot spot in the upper mantle, has produced three colossal eruptions during the last two million years, two of which are among the most violent in earth's history. Earthquakes, ground swelling, and increasing thermal activity suggest that the large body of hot rock beneath Yellowstone may produce another world-class eruption. The three previous climactic outbursts were spaced about 600,000 years apart. The same length of time has elapsed since the last great event, so a fourth paroxysm may be imminent. Because no volcano has erupted on so large a scale in historic time, we do not know what warning signs precede such an eruption. Another climactic outburst will probably begin with numerous preliminary earthquakes, ground deformation, and greatly accelerated geyser and steam vent activity.

Some volcanic danger signals are now evident in the Mono Lake-Long Valley region of east-central California. The swarms of earthquakes that have shaken the area since 1978 may reflect the fracturing of subterranean rock as tongues of magma are injected into the upper crust. If an eruption occurs, it is likely to resemble activity like that of about A.D. 1350, when a chain of separate erupting vents was blasted open at both the north and south ends of the Mono-Inyo Craters. That eruptive cycle produced widespread ashfalls over central California and western Nevada, as well as pyroclastic flows and a long series of thick streams of rhyolite lava. A late twentieth century repetition of the most recent events could seriously endanger nearby communities such as Mammoth Lakes or other settlements on U.S. Highway 395, eastern California's main thoroughfare.

Many of the recent earthquakes center beneath Long Valley, a vast collapse depression created by the same kind of catastrophic eruption that formed the Yellowstone caldera. About 700,000 years ago, a volcano at the site expelled enormous quantities of incandescent ash, forming pyroclastic flows that traveled at speeds exceeding 100 miles per hour. So great was the volume of hot ash and so high its velocity, that one arm of the ashflow surmounted the steep eastern face of the Sierra Nevada, a granite barrier thousands of feet above the erupting vents. Overtopping the mountain crest, the ashflows rushed west down the San Joaquin drainage, perhaps as far as the Central Valley of California.

Another arm of the pyroclastic flows traveled at least forty miles south down the Owens Valley to the present site of Bishop. At least 580 square miles of central California and western Nevada were buried under a thick layer of ash, the Bishop tuff, while the vertically ejected eruption cloud deposited enough ash in the Midwest that some

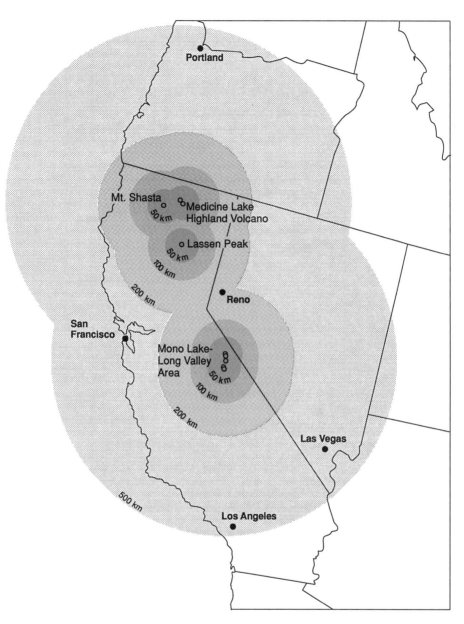

Map showing volcanoes in California that may produce large to extremely large eruptions of tephra, airborne rock fragments. The outermost circles (light pink) represent areas that could be affected by significant ashfall during an eruption as large as that which deposited the Bishop tuff and formed the Long Valley caldera about 700,000 years ago. A much smaller eruption, comparable to that which formed Crater Lake about 6,900 years ago, would deposit ash more than a foot deep sixty miles downwind from the volcanic vent. Lighter ashfalls would extend for hundreds of miles, depending upon wind strength and direction. —After C. Dan Miller, 1989, U.S. Geological Survey Bulletin 1847

of it remains today in central Nebraska. The volume of fresh magma erupted during a few hours or days, 140 cubic miles, compares with that thrown out by the Yellowstone volcano. During the ash flows, the roof of the Long Valley magma chamber collapsed, forming an oval caldera eleven by eighteen miles in diameter.

The Yellowstone and Long Valley calderas are only two of the many similar features that pockmark the western landscape in Nevada, California, Colorado, Idaho, Wyoming, and New Mexico. All are considerably larger than the collapse basins formed by the destruction of the former summits of Cascade stratovolcanoes, such as Mazama or Newberry, or the Aleutian Range volcanoes, such as Katmai, Okmok, or Aniakchak.

The vast size of some of these western calderas, which must be viewed from the air to be appreciated, suggests the enormous volume of material they ejected. The Valles caldera in the Jemez Mountains about thirty-five miles northwest of Santa Fe, New Mexico measures twelve by fifteen miles. The nearby Toledo caldera produced about fifty cubic miles of ash flows about 1.3 million years ago and then was itself buried by even more voluminous eruptions of the Valles caldera about 300,000 years later. Creede caldera in the central San Juan Mountains of Colorado is about fourteen miles in diameter. Nevada's Timber Mountain caldera is still larger, about eighteen miles wide and twenty miles long. All formed during the violent eruptions of phenomenally large quantities of rhyolite magma.

Fortunately for humanity, no volcanic eruptions on this scale have occurred anywhere in the world since long before the dawn of civilization. Nonetheless, the large number and wide distribution in time and space of these catastrophic events suggests that, however rare, they will occur again. Another outpouring of incandescent rhyolitic ash equivalent in volume and power to the outbursts that opened the Yellowstone and Long Valley calderas, 600,000 and 700,000 years ago, respectively, may not happen for tens of thousands of years. Nature gives no guarantee, however, that it might not occur much sooner.

Part III:
Ice and Fire, Myth and Reality

Chapter 25

A LOOK BACK INTO THE ICE AGE:
Glacier Bay, Alaska

A sharp cracking noise like artillery fire drew all eyes toward the blue-silver wall of ice towering above our ship. Then an enormous slab of ice the size of a twenty-story building broke loose from the glacier's cliff-like front and plunged into the deep cold waters of Glacier Bay. Waves rushing outward from the newly created iceberg threatened to capsize the six kayaks that ventured dangerously near the crumbling ice mass.

Massive rivers of ice thrusting their snouts into Alaska's Glacier Bay provide one of nature's greatest spectacles. Every day the bay's sixteen tidewater glaciers, glaciers that flow directly into the ocean, generate hundreds of such icebergs and ice floes. The day we watched, from the relative safety of a large cruise ship, the kayaks escaped unscathed, but many other small vessels have been swamped and their occupants drowned by cascading ice and surging sea.

Flowing from Alaska's mountainous interior, a massive tidewater glacier terminates in a steep wall of ice at Glacier Bay. Such rivers of ice have played a major role in sculpturing the North American landscape.
—U.S. Geological Survey, Austin Post photo

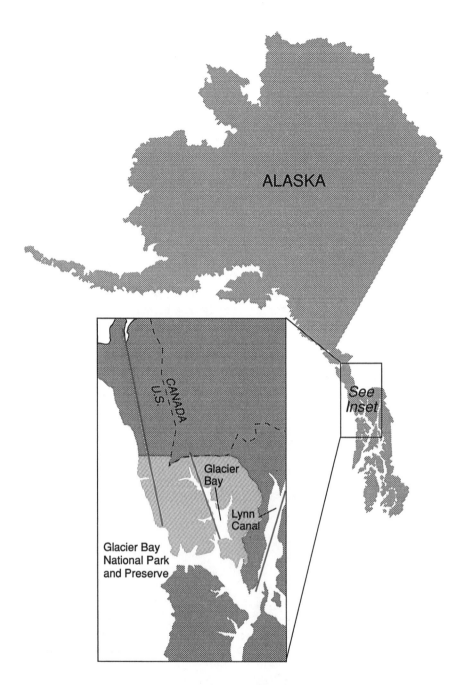

Map showing location of Glacier Bay in the Alaska panhandle. Note that the region is crossed by numerous earthquake faults.

Floating glaciers constantly calve icebergs into the waters of Glacier Bay National Park. Many are large enough to scuttle good-sized ships that by some unlucky chance collide with them. Like many of nature's grandest shows, the confrontation of ice stream and ocean offers a combination of unforgettable beauty and potential danger. Although the view from aboard ship is awesome, a cruise through Glacier Bay reveals only part of the picture. Calving icebergs is merely one aspect of the glaciers' activity. To get an idea of the whole process of glaciation the visitor must fly inland and find the glaciers' place of origin.

Soaring in a small plane over the jagged peaks that rise steeply above Glacier Bay, the traveler enters a pristine world of naked rock and ice. Travelers do not simply venture into a rugged mountainous terrain untouched by man, they journey back in time to see what much of the world was like during the Pleistocene epoch, the series of Ice Ages that lasted from about 1.8 million years to 12,000 or 10,000 years ago.

The Fairweather and St. Elias Ranges rise suddenly above the coastline to elevations of more than three miles above sea level. These highest coastal mountains in the world are also the most heavily glaciated. Mount St. Elias reaches an elevation of 18,008 feet and Mount Fairweather stands at 15,300 feet. Their size and height makes these mountains a climatic barrier that intercepts moisture-laden winds from the Pacific. They receive abundant precipitation, in places more than 200 inches of snow a year, furnishing the basic ingredient of glaciers.

As winter storms deposit layers of new snow, its weight forces the air out of the snowpack. The tightly compacted snow finally becomes dense layers of ice that eventually begin to move downhill under the pull of gravity. All but the loftiest peaks and ridges around Glacier Bay are smothered in moving ice.

A flight over the Fairweather and St. Elias Ranges shows that the sixteen glaciers flowing into the bay are merely arms of an immense icefield that engulfs southeastern Alaska's interior highlands. Ice sheets cling to all but the steepest pinnacles and fill broad valleys to depths of thousands of feet. Mountain glaciers sweep down adjacent valleys like gigantic white freeways and merge to form immense icesheets many miles wide. The glaciers even have marked lanes, long parallel bands of white, with dark strips running the length of the glacier. The dark strips are rock debris that drops onto the edge of the glacier; when two glaciers merge, the bands of debris along their edges become a dark stripe down the ice.

Ice flows calved from the front of the Plateau Glacier litter a cove of Alaska's Glacier Bay. The ice-shrouded Fairweather Mountains in the background provide an image of what much of North America looked like during the Pleistocene Epoch or Ice Age. —U.S. Geological Survey, Austin Post photo

The coastal ice fields near Glacier Bay now mantle thousands of square miles, but they were much larger 200 years ago when Europeans first entered the area. In 1794, when Captain George Vancouver sailed past what is now Glacier Bay, it was only a small dent in an unbroken mass of ice that sealed the entire inlet. The region was still in the grip of the Little Ice Age, a period of globally cooler and wetter weather that began about 3,500 years ago and lasted until about the middle of the nineteenth century. World climate has since turned milder and Glacier Bay's ice streams have retreated. When John Muir first visited the area in 1879, he found that the ice sheet that blocked the bay in 1794 had withdrawn about forty-eight miles.

As the climate warmed, the Glacier Bay ice streams staged the fastest retreat of any glaciers known to science. Some withdrew upvalley from the shoreline, leaving bedrock exposed to air for the first time in millennia. By 1916 the front of the Grand Pacific Glacier stood sixty-five miles from the mouth of Glacier Bay. In the 1920s its terminus withdrew 1.2 miles across the Canadian boundary. By 1948 the Grand Pacific had pushed its way back across the United States border. Today it discharges icebergs into the sea at Tarr Inlet from a terminus nearly four miles across and up to 260 feet high.

Although the Grand Pacific Glacier continues to advance, the Muir Glacier shows no sign of halting a retreat begun well over a century ago. Since the time when Muir lived in a one-room cabin near its calving snout, the glacier has melted back twenty-five miles. Many other glaciers are also retreating, but a number are either holding their own or even surging forward.

What causes some tidewater glaciers to advance while others withdraw? Why do they not respond to today's warmer climate in the same way? The explanation of the glaciers' lack of coordination involves the positions of their fronts and the relative elevation of the inland ice fields in which they originated. When a glacier front extends into an area where the rate of melting exceeds the volume of ice flowing down from the mountains, it typically dumps its accumulated load of rock debris to form a terminal moraine. The glaciers extending into Glacier Bay rest their snouts on top of these ridge-like moraines. When unusually rapid melting causes the glacier front to withdraw from the terminal moraine to deep water behind it, the glacier begins to disintegrate until it reaches shallow water. The Muir Glacier will probably retreat until it reaches shallow waters in which it can stabilize. At present the advancing glaciers head in alpine ice fields that have elevations averaging about 6,500 feet. The immense snowfall that annually blankets these ice caps keeps the glaciers well nourished. The ice streams now retreating from Glacier Bay originate at much lower elevations and do not benefit from the plentiful snows that feed their growing neighbors.

Glaciers are highly sensitive indicators of climatic change. Many scientists fear that global warming, compounded by industrial society's injection of carbon dioxide into the atmosphere, will hasten the dreaded greenhouse effect. Rising temperatures could melt the polar ice caps and raise sea levels, flooding the world's coastlines.

Despite anxiety about global warming, earth's present warm climate may be a brief interval between Ice Ages. The long cold night of the Pleistocene epoch was interrupted by periods of warming when continental ice sheets disappeared. Some of these ice-free episodes lasted tens of thousands of years, much longer than the Holocene epoch, the most recent 10,000 years.

Events that initiated the Ice Ages 1.8 million years ago may happen again. Dust from a meteorite impact, a close-encounter with a comet, or unusually large volcanic eruptions, could envelop the earth, reflecting sunlight and lowering global temperatures. After only a few decades in which more snow falls in winter than melts in summer, large continental ice sheets could form, further chilling the climate. During the Pleistocene epoch such ice sheets spread through northern Europe and Asia, and virtually all of Canada, bulldozing their way into the northern United States. Simultaneously, thick ice caps accumulated along mountain belts, engulfing all but the loftiest summits in masses of grinding ice. In some places, alpine glaciers extended tens

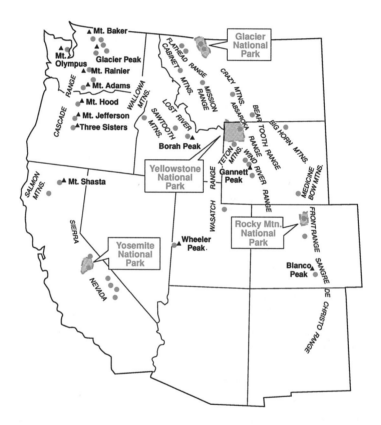

Glaciers, shown here as red dots, are now largely confined to high peaks in the Cascades, Olympics, Rockies, and southern Sierra Nevada. The largest mantle Mount Rainier and the Northern Cascades in Washington State. Most U.S. glaciers are not shrunken relics from the Ice Age, but formed during a neoglaciation that started about 3,500 years ago.

of miles downvalley to merge with lowland ice sheets descending from the north. With seventeen million cubic miles of earth's surface water locked up in ice, Pleistocene oceans dropped 300 to 400 feet below their present levels. These drastic falls in sea level exposed land bridges, one of which linked Asia and North America, permitting Old World inhabitants to set foot on our continent for the first time.

During the most recent 10,000 years, briefer episodes of world cooling caused glaciers to reform. About 3,500 years ago, new glaciers appeared in the Rockies, Cascades, Sierras, and other mountains of the western United States. Another Little Ice Age, beginning in the fourteenth century and ending about 100 years ago, also caused glaciers to expand. Pioneers settling the Pacific Northwest and the northern Rocky Mountain states during the 1800s found significantly larger glaciers than those we have today.

The present warming trend may eventually destroy most of our remaining glaciers, or it may be reversed by a return to Ice Age conditions. Whether the future brings increased heat or cold, expanding deserts or land-smothering ice sheets, glaciers may provide nature's first warning of impending change.

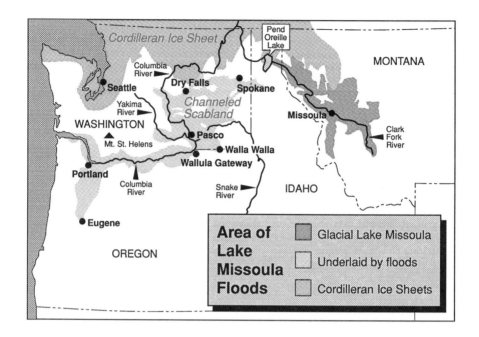

The world's largest floods poured from glacial Lake Missoula across Washington, down the Columbia River Gorge, and into the Pacific Ocean, scouring much of the Columbia Plateau down to bedrock and creating the Channeled Scabland. As many as forty to seventy separate floods occurred between about 15,000 and 12,000 years ago. Note that the floods also inundated the Willamette Valley as far south as Eugene. Deposits from the retreating Cordilleran Ice Sheet now mantle the flood path between the Clark Fork River and Spokane.

Chapter 26

GLACIAL LAKE MISSOULA:
The World's Largest Flood

"You Greeks [and Americans] are all children....you have no belief rooted in old tradition and no knowledge hoary with age....You remember only one deluge, though there have been many...."

> —Egyptian Priest to Solon, the Athenian lawgiver, discussing the many catastrophes "by fire and water" that have repeatedly devastated humankind. From Plato, the *Timaeus* (introducing the Atlantis myth).

When early nineteenth-century missionaries to the Pacific Northwest learned that several native American tribes preserved legends of a Great Flood, it seemed to many believers a welcome confirmation of the Biblical deluge. The science of geology was in its infancy, and many Victorian scientists interpreted earth's history as a series of catastrophic upheavals—relatively short-lived, violent phenomena that elevated mountain ranges, threw down valleys, and laid out the thick deposits of sediments now exposed in deep canyons. Noah's flood was typically seen as only the last of these events.

Opposed to these catastrophists were the uniformitarians, who argued that virtually all geologic features can be explained by the same slow processes of eruption, erosion, sedimentation, and ground movement we see at work today. Careful study of the rocks ultimately vindicated the uniformitarian position, which was further supported by the discovery of our planet's enormous age. The concept of deep time, the 4.6 billion years during which the earth developed into its present state, granted the chronological space needed to account for everything from the Grand Canyon to marine fossils in the Alps.

The Deschutes River empties into the Columbia River at the eastern end of the gorge. Note the barren rock in the foreground. All topsoil was removed by the gigantic Lake Missoula floods. —Delano Horizons Inc., Leonard Delano photo

So diligently had the scientific community labored to escape from the inherited traditions of a six-day creation and Noachian deluge, that eventually it became understandably reluctant to view almost any formation as other than the product of long shaping by natural forces operating at the normal rate. Ironically, it was necessary to revert to a kind of limited catastrophism to understand one of the largest and most spectacular terrains in the West, the 2,000 square-mile Channeled Scabland in east-central Washington. For decades most geologists vigorously resisted the notion, first presented in the 1920s by J H. Bretz, that this vast tract of deeply scoured bedrock was formed not during eons of ordinary erosion but in mere days or weeks by floods of almost unimaginable magnitude.

Only from the air can one easily appreciate the immense channels carved when cataclysmic floods roared hundreds of feet deep across northern Idaho and the lava plains of eastern Washington, on their way down the Columbia River to the Pacific. The floodwaters peeled

196

away thick layers of basaltic lavas, excavating an intricate network of channels, one as deep as 900 feet, and leaving behind colossal gravel bars with giant dunes sculpturing their surfaces, fossil wave-crests up to twenty feet high and 300 feet from crest to crest. Some high ridges were left as islands, their up-stream ends whittled to prowlike points. The coulees of the Columbia Plateau then thundered with waterfalls and cataracts of immense proportions. The largest was Dry Falls in the southern part of Grand Coulee. Four hundred feet high and almost three miles wide, it is two and one-half times as high and five times as wide as Niagara Falls today. For brief periods, Dry Falls carried an immeasurably greater volume of water.

The Lake Missoula Floods scoured eastern Washington's Scabland down to bedrock. During the series of catastrophic floods that repeatedly swept across the Columbia Plateau during the late Ice Age, Dry Falls, in the center, carried more water than today's Niagara Falls. Note the contrast between the naked rock in the center of the photo and the irrigated terrain in the left background, representing original soils untouched by the cataclysm. —Delano Horizxons, Inc., Leonard Delano photo

The catastrophic Lake Missoula floods carried this boulder hundreds of miles from its place of origin and deposited it as an erratic along the Columbia River in eastern Washington.
—Delano Horizons, Inc., Leonard Delano photo

The origin of the floodwaters lay in northwestern Montana where, about 16,000 years ago, a huge glacial lake, Lake Missoula, had formed. At that time a lobe of the continental ice sheet advancing southward from Canada blocked the Clark Fork Valley near Pend Oreille Lake in the Idaho panhandle, creating an ice dam thousands of feet thick and impounding a body of water that stretched southeastward about 250 miles. When the ice dam failed, it released the entire glacial reservoir, perhaps 500 cubic miles of water, which rushed westward to emerge on the Columbia Plateau near the site of Spokane. From there the floods poured west and southwest, stripping away the soil and rapidly eroding the basaltic bedrock. Erratic blocks of ancient sedimentary rocks were carried hundreds of miles from their source and deposited along the Columbia River where they can be seen stranded along the walls of the Columbia Gorge. Rafted by floating blocks of glacial ice, some of these erratics were swept as far as the Willamette Valley in western Oregon. Others, originally plucked from glaciated peaks high in the Rockies, were eventually swept down the Columbia into the Pacific Ocean, a distance of at least 430 miles.

Pouring out at perhaps 9.5 cubic miles per hour for at least forty hours, a discharge estimated to be ten times the combined present flow of all the world's rivers, the flood temporarily ponded in the Pasco Basin, the lowest point in its journey over the Columbia Plateau, briefly forming a large lake. Because the only exit from the Pasco Basin, the Wallula Gap on the Washington-Oregon border, was too narrow to allow such an enormous quantity of water to escape readily, the hydraulically dammed floodwaters flowed upsteam into the Yakima, Walla Walla, and Snake River valleys.

Cresting at about 1,200 feet at the Wallula Gateway, the flood spread out to cover an area of 1,300 square miles in the Umatilla Basin, temporarily creating another sizable Pleistocene lake. Streaming down the Columbia Gorge, widening the valley and removing soil up to elevations of 1,000 feet as far as The Dalles, Oregon, the turbulent mass churned through The Dalles Basin, carving channels and depressions, some up to 225 feet below sea level, over a 100-square-mile area.

By the time it reached the Hood River's confluence with the Columbia, the floods still rose approximately 900 feet, cutting away the lower courses of tributary streams that entered the Columbia Gorge, leaving them hanging 400 feet above the present river surface. Entering the Portland area, the floods spread southward into the

This view of the Columbia River below the Wallula Gap reveals the enormous eroding power of the Lake Missoula floods. The floodwaters not only scoured the land down to bedrock, they peeled away large sections of the underlying basalt, leaving distinctive terraces and mesas along the right-side riverbank.
—Delano Horizons, Inc., Leonard Delano photo

Looking eastward over Rowena Point. The Lake Missoula floods repeatedly swept across this scene, widening and steepening the Columbia River Gorge. Note the layers of basaltic lava on the northern side of the gorge and the deeply scoured depression in the foreground.
—Delano Horizons Inc., Leonard Delano photo

Willamette Valley as far south as Eugene, inundating 3,000 square miles as deeply as 350 feet.

Perhaps as astounding as the erosive force and magnitude of the deluge from glacial Lake Missoula is the number of times that such floods recurred. Field research by Richard B. Waitt, Jr., of the U.S. Geological Survey, Dave Alt of the University of Montana, and others, concludes that the ice dam formed no fewer than forty times, only to be destroyed and to release its prodigious quantities of water. Sufficient time elapsed between some of the floods for wind-blown soil to form atop flood deposits in the Walla Walla and lower Yakima River valleys, which are now underlaid by forty or more thin layers of flood-deposited silt, sand, and gravel. The intra-flood deposits also include a two-part ashfall from St. Helens that was erupted about 13,000 years ago.

The repeated emptying of Lake Missoula happened when the ice dam floated and broke. As the lake became approximately nine-tenths as deep as the ice dam was thick, the glacier floated and released the impounded water. The mechanism is probably the same as that of modern but much smaller ice-dammed lakes, such as those in Alaska and Iceland. Meltwater flowing into Lake Missoula from the continental ice sheet on the north, as well as the influx from various rivers, probably refilled the lake within a few decades. Recent investigations indicate that each successive filling took less time than the one before, suggesting that the ice dams were progressively thinner and released increasingly shallow lakes.

It is not certain that native Americans then inhabited the Pacific Northwest. The oldest Indian artifacts found in eastern Washington are about 11,000 years old. Certainly few people who might then have lived along the flood-course would have survived to relate their interpretation of the gods' wrath in thus drowning the world. Yet survivors in areas bordering the flood or atop islands of high ground, isolated arks of safety, would undoubtedly have been sufficiently impressed to pass on oral histories of a great flood. Tribal memories of one or more of these watery cataclysms may have inspired the accounts related to Northwest missionaries. Although transmission of oral history for 12,000 years may seem improbable, some anthropologists point out that if events similar to those in the tradition, such as large floods along the Columbia River, remain part of the tribal experience, the collective memory will persist.

Nature consistently eludes attempts to pin her behavior down to any inflexible scientific law. Most of her earthly work operates at the observed, normal pace, gradually raising mountains by almost imperceptible earthquakes and carving valleys by routine stream-cutting. But the element of the unexpected, of chaos, can interrupt the ostensibly fixed geologic processes with a sudden violence that challenges our sense of natural order. Such were the Lake Missoula floods.

Looking westward down the Columbia River toward the snowcapped Cascade Range.
—Delano Horizons Inc., Leonard Delano photo

The Emmons Glacier flows from the summit ice fields of Mount Rainier, elevation 14,410 feet, down the volcano's steep eastern flank into the White River valley. During the last 7,000 years, numerous eruptions have generated colossal floods and mudflows that have swept as far as sixty-five miles downstream from the summit, inundating areas where 60,000 people now live. Yakima Park and the Sunrise Visitor Center appear on the right.
—U.S. Geological Survey, Austin Post photo

Chapter 27

FIRE UNDER ICE:
The Steam Caves at
Mount Rainier National Park

Volcanic fire and glacial ice are natural enemies. Eruptions at glaciated volcanoes typically destroy ice fields, as they did in 1980 when seventy percent of St. Helens' ice cover was demolished. During long dormant intervals, glaciers gain the upper hand, cutting deeply into volcanic cones and eventually reducing them to rubble. Only rarely do these competing forces of heat and cold operate in perfect balance to create a phenomenon such as the steam caves at Mount Rainier National Park.

Few of the park's two million annual visitors are aware that Rainier not only supports the United States' largest glacier fields south of Alaska but also contains the nation's most extensive steam caves. Located inside Rainier's two ice-filled summit craters, these caves form a labyrinth of tunnels and vaulted chambers about one and a half miles in total length. Similar interconnecting ice tunnels are known to exist at only two other mountains in North America, Alaska's Mount Wrangell and Rainier's Cascade neighbor, Mount Baker. Their creation depends on an unusual combination of factors that nature almost never brings together in one place. The cave-making recipe calls for a steady emission of volcanic gas and heat, a heavy annual snowfall at an elevation high enough to keep it from melting during the summer, and a bowl-shaped crater to hold the snow.

Snow accumulating yearly in Rainier's summit craters is com-pacted and compressed into a dense form of ice called firn, a substance midway between ordinary ice and the denser crystalline ice compos-ing glaciers. Heat rising from numerous fumaroles along the inner crater walls melts out chambers between the rocky walls and the overlying icepack. Circulating currents of warm air then melt additional openings in the firn ice, eventually connecting the individual chambers and, in the larger of Rainier's two craters, forming a continuous passageway that extends two-thirds of the way around the crater's interior.

To maintain the cave system, the elements of fire under ice must remain in equilibrium. Enough snow must fill the crater each year to replace that melted from below. If too much volcanic heat is discharged, the crater's ice pack will melt away entirely and the caves will vanish along with the snows of yesteryear. If too little heat is produced, the ice, replenished annually by winter snowstorms, will

Heat and steam emission from Rainier's ice-filled summit craters have melted a series of interconnecting chambers and passageways between the inner crater walls and the ice pack, over 14,000 feet above sea level. Baker and Alaska's Wrangell are the only two other volcanoes in North America known to have a similar cave system. —Modified from Kiver, 1971 and 1975

expand, pushing against the enclosing crater walls and smothering the present caverns in solid firn ice.

Rainier's steam cave system has remained relatively stable since it was first discovered in 1870, when two climbers making the first official ascent took refuge in ice caverns along the crater rim. Like many a benighted climber since, Hazard Stevens and P. B. Van Trump spent a miserable time, half-scalded by steam jets on one side and frozen by gale-force winds on the other. Without lanterns, they did not explore the deeper tunnels that extend at least 360 feet below the ice surface.

When Hazard and Van Trump made their discovery, the present caves had probably existed for only a few years. During the mid-nineteenth century a meltwater lake may have occupied Rainier's crater, if the report of a Yakima Indian named Saluskin is reliable. Recorded when Saluskin was an old man, his account states that in 1855 he had guided two "King George men"—the Indians' term for Caucasians—to a base camp on Rainier's east flank, from which they ascended the Emmons Glacier to the summit. Upon returning, the two adventurers told Saluskin that there was "ice all over top, lake in center and smoke or steam coming out all around like sweat house." Although some historians question the accuracy of Saluskin's anecdote, others note that the only way for anyone to have known about Rainier's intense thermal activity was to have reached the crater rim.

The presence of a steaming lake at Rainier's summit in 1855 seems plausible considering that eyewitnesses recorded numerous small eruptions of the volcano during the 1840s and 1850s. Increased heat emission during this eruptive activity would have transformed the crater icefill into a sizeable body of standing meltwater. Even today a meltwater pool about 120 feet long and eighteen feet deep exists in Rainier's smaller west crater. Concealed by an overarching canopy of firn ice more than 14,000 feet above sea level, it is the highest such pond on the continent.

In 1975 Rainier's northern sibling, Mount Baker, demonstrated how quickly a crater lake can form, radically altering the volcano's steam cave system. That March witnessed a sudden and dramatic increase in the heat and volcanic gas issuing from Sherman Crater, a large funnel-shaped vent located about 1,000 feet below Baker's heavily glaciated summit. Spewing clouds of steam and ash, Baker's numerous fumaroles began ejecting sulphur at the rate of 21,000 pounds per hour. Reacting like snow balls on a frying pan, the crater icepack rapidly broke up, exposing parts of the crater floor and forming a shallow acidic lake that measured about 164 by 230 feet.

Less than a year before this thermal event, the author had joined the first expedition to survey and map the maze of tunnels and caverns in Sherman Crater. At that time one could enter the cave system through openings in the ice pack along the crater's western rim and follow a lofty-ceilinged passageway all the way across the crater floor, about a third of a mile, emerging at a huge gap in the crater's eastern wall where pulsating steam jets shot columns of vapor hundreds of feet into the air.

Gene Kiver, a geologist and spelunker from Eastern Washington University, led our party into the Stygian darkness of the unknown. We were probably the first to investigate Baker's steamy caverns in depth. They lie off the main climbing routes and are so permeated with volcanic gases, including carbon dioxide and carbon monoxide, that would-be explorers must bring gas masks.

In 1990, Baker's steam and sulphur production continued to operate significantly above pre-1975 levels, although thermal action has declined somewhat from the 1975-76 maximums. In 1977 an ice avalanche from the north summit buried the former lake site so that if a lake continues to exist, it is now subglacial.

At Rainier and Baker volcanic fire and glacial ice peacefully co-exist, but their long-term association may prepare the way for future catastrophe. A new study of Rainier's hot springs and fumaroles suggests that acidic fluids percolating through the cone have extensively altered the interior rock, changing it into soft clayey material. The present summit cone, built about 2,000 years ago, sits atop the volcano's heat-decayed core, a friable surface "analogous to a greased bowl." Injection of fresh magma into the young cone, already tilted toward the northeast, may trigger its collapse and create a huge avalanche like that on St. Helens in 1980. Sudden removal of the summit cone could uncork magma and superheated water inside the volcano, releasing a laterally-directed explosion or pyroclastic surge that would sweep over neighboring peaks and ridgetops, destroying all in its path.

Water-saturated rock in the avalanche would soon transform it into a mudflow inundating the White River valley. Additional melt-water from breakup of the Emmons and Winthrop glaciers and from Rainier's east crater, which now contains snow and ice equivalent to a billion gallons of water, could further mobilize the avalanche, allowing it to travel many miles downvalley and spread into the heavily populated Puget lowland. A large-volume mudflow from Rainier could claim more lives than that at Columbia's Nevado del Ruiz, which killed 25,000 people in 1985. In a relatively small eruption, the Columbian volcano generated floods and mudflows that entombed an entire city with most of its inhabitants.

With its vast bulk sitting high astride intersecting ridges and deep valleys, Rainier's heat-decayed cone is potentially dangerous even when the volcano is quiet. Although most of its great avalanches and mudflows were eruption-related, some of the largest had other causes. About 2,800 years ago an enormous rocky mass peeled away from Rainier's western face, creating a mudflow at least 1,000 feet thick as its streamed down the Tahoma Creek and South Puyallup River valleys. Containing a high proportion of hydrothermally altered rock similar to that being formed at Rainier's summit today, this mudflow opened a large scar, the Sunset Amphitheater, and exposed the volcano's permeable core. Another exceptionally large mudflow, the Electron, originated in the same area about 600 years ago and traveled at least forty miles into the Puget lowland. Both catastrophic landslides may have been triggered by high magnitude earthquakes on the Cascadia subduction zone. During future great earthquakes in the Pacific Northwest, inhabitants must cope not only with falling buildings but also with collapsing volcanic cones that may produce landslides and mudflows extending beyond the Cascade Range into thickly settled areas.

Besides posing volcanic and seismic hazards, Rainier also produces sudden rockfalls that can affect campers, hikers, and climbers. In 1963 approximately 14 million cubic yards of rock avalanched from Little Tahoma Peak on Rainier's east side. Perhaps initiated by a steam explosion, the Little Tahoma rockfalls traveled four miles down the Emmons Glacier into the White River valley, stopping just short of a campground. In 1989 another large rockfall partly buried Winthrop Glacier on the northeast flank. Geologists fear that disastrous avalanches will occur without warning, some initiating mudflows that can invade communities many miles downvalley from the volcano.

Rainier's heat and steam emission have produced one of nature's rarest phenomena, the steam cave system that lies hidden inside the summit. The same hydrothermal process has also chemically altered the volcano's interior, over millennia transforming solid rock into slippery clay vulnerable to sliding and collapse. The present delicate balance between fire and ice will end during the next eruption or great earthquake, perhaps in a cataclysmic disruption of the cone that will change America's most majestic peak into an agent of destruction.

Looking northeast across the toe of the Bonneville Slide, which temporarily dammed the Columbia River about A.D. 1100 and may have inspired the Bridge of the Gods legend. Today's steel "Bridge of the Gods" spans the river at Cascade Locks near the top center, while Bonneville Dam appears in the central foreground. —Delano Horizons, Inc., Leonard Delano photo

Chapter 28

INDIAN MYTH AND GEOLOGIC REALITY: The Bridge of the Gods

Systematic study of the West's geology began about 1879 when the United States Geological Survey was founded. But for thousands of years before scientists began to unravel the complex evolution of the western landscape, American Indians preserved an oral tradition of some major geologic events. The new discipline of geomythology studies ancient myths that reflect natural events such as earthquakes and volcanic eruptions.

In a prescientific age, the native people of North America interpreted geologic processes as part of a unified view of the world that blended physical and spiritual components. Mountains were not simply inert piles of rock; they were living presences with distinctive personalities. Rivers, lakes, and other topographical features were inhabited not only by animals, who were also kin to humankind, but also by unseen beings who could manifest their power through events that today we are taught to regard as the impersonal working of geophysical laws. In native traditions heaven, earth, and human consciousness were bound together in a rhythm of life, a sense of wholeness.

One of the most attractive legends illustrating the bond connecting earth and its inhabitants concerns the Bridge of the Gods, which reputedly once spanned the Columbia River near the present site of Cascade Locks, Oregon. Numerous versions of the story exist, and it is probably impossible now to disentangle authentic legends from later embellishment and Caucasian reinterpretation. According to one account of the Klickitat tribe, long before white people appeared

Looking southeastward across the Bonneville Slide (right center). Triggered about A.D. 1100 by the collapse of the high ridges on the northern side of the Columbia, the slide formed a dam over a mile wide and 200 feet high. While it lasted, this tahmahnawis *formation served the Indians as a "bridge of the gods."* —Leonard Delano photo

on the scene, native tribes could use the bridge to cross the Columbia dry-shod. But when the tribes became greedy and quarrelsome, Coyote or the Great Spirit took steps that eventually led to the bridge's destruction. First he caused all the fires in their lodges to go out. Only the fire maintained by Loowit, an aged lady who avoided the violence that divided her people, remained burning so that all her neighbors had to come to her to relight their campfires. When the Great Spirit asked Loowit to name a reward for her peaceful sharing, she instantly demanded youth and beauty. Transformed into a lovely young woman, Loowit inadvertently rekindled the fires of war. For she now attracted the admiration of two great chiefs, Phato who ruled over the northern, Washington state, side of the Columbia and Wyeast who led the Willamette people south of the river.

Phato and Wyeast contended fiercely for Loowit's favor, hurling red-hot boulders at each other and causing the earth to tremble. Angry at the chiefs' destructiveness, the Great Spirit separated them by destroying the bridge, its giant fragments creating the Cascades of the Columbia, the cataracts for which the neighboring Cascade Range is named. The Spirit also changed the three principals in the love triangle into mountains. Wyeast became the lofty pyramid white people call Mount Hood. Phato became broad-shouldered Mount Adams, while Loowit became the smoothly youthful cone of Mount St. Helens.

Even after their metamorphosis into high peaks, the trio continued to blaze with passion, spewing smoke and flame into the air. Being the youngest, Loowit remained infatuated the longest. After her suitors eventually fell silent under mantles of ice and snow, Loowit still burned, becoming known as Tah-one-lat-clah—"fire mountain."

Behind the Klickitat legend lie an impressive number of geologic facts. The correlation between story and reality is clearer once certain misconceptions are corrected. While most nineteenth century story-tellers envisioned the Bridge of the Gods as a high natural stone arch spanning the Columbia Gorge from rim to rim, what appear to be the earliest forms of the tale make no such claim. The natives of the region did not have a word for bridge in their language and stated only that the river was dammed so that many people at once could walk across the obstruction without getting their feet wet.

Considerable evidence suggests that such a massive earth dam did exist, temporarily blocking the Columbia's flow. On his party's return trip through the gorge in 1806, Meriwether Lewis marvelled at finding a drowned forest upriver from the Columbia Cascades. Seeing thousands of tree trunks exposed at low tide or standing erect in ten to thirty feet of water, Lewis estimated that the river had been dammed only twenty years before. A 1958 radiocarbon dating of wood from the ghost forest, however, indicated that the pine snags had been killed about A.D. 1260. Later dating of wood samples taken from the Bonneville Slide itself revealed that the slide had occurred about A.D. 1100.

The largest of many landslides to affect the north bank of the Columbia River since the Lake Missoula floods thundered through, the Bonneville Slide was the result of a peculiar combination of geologic circumstances. The series of Lake Missoula floods widened the Columbia Gorge and steepened its walls, leaving them unstable and subject to slumping and sliding. The gorge walls are particularly unstable because the upper 1,500 feet are composed of dense Columbia River basalt flows while the lower 1,000 feet consist of easily eroded rock. This bottom layer, the Eagle Creek Formation, is formed largely of old volcanic ash beds and mud flow deposits. The whole sequence is tilted gently southward so that the upper basalt flows on the north side of the river slide over the water-saturated Eagle Creek material, in the words of one geologist, as if they were "a stack of heavily buttered pancakes on a tilted platter."

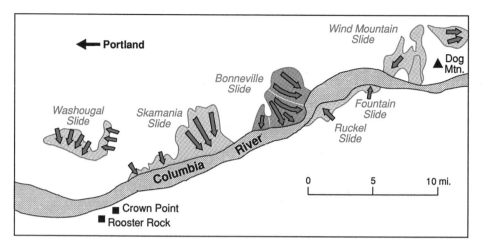

During recent geologic time, numerous large landslides have avalanched down the north wall of the Columbia River Gorge, the largest of which is the Bonneville Slide of about A.D. 1100. The north side is particularly prone to sliding because the lower parts of the gorge wall are composed of slippery clays and sedimentary rock overlain by massive lava flows from the Columbia Plateau and the Cascade volcanoes. Uplift of the Cascade Range tilts the formation southward, causing the heavy basalt rocks periodically to topple into the gorge.

The event that triggered the massive Bonneville Slide may have been a high magnitude earthquake, perhaps on the Cascadia subduction zone. Whatever the mechanism, nearly half a cubic mile of rock broke loose at or near the summits of Table Mountain and Greenleaf Peak, which stand about three miles north of the river. A series of catastrophic landslides buried nearly fourteen square miles of the Columbia Gorge and forced the river more than a mile south to its present course. This created a natural dam more than 200 feet high.

We cannot know how long the slide blocked the Columbia, but it may have taken years for the impounded water to overtop the dam and wash most of it away. While the dam lasted, it served as a land bridge that permitted the Klickitats, Willamettes, and other tribes to enjoy easy access to each other's territory. Even after the river cut through the toe of the slide, great basalt slabs partly blocked the channel, forming the Columbia Cascades.

What of the mountain participants in the bridge's fall? The legend bearers correctly assessed the volcanic nature of the three glacier-draped peaks known as the Guardians of the Columbia. Although the least active of the three, Mount Adams has repeatedly erupted lava flows during recent geologic time. Hood produced at least three major eruptive cycles within the last 1,700 years. Between about 1760 and 1810, Hood was intermittently active, erecting the still-steaming lava dome called Crater Rock and sending mud flows and floods down-valley as far as the Columbia. Old Wyeast's most recent flare-ups occurred during the period of historic observation, in 1859 and 1865-66.

Mount St. Helens (Tah-one-lat-clah) has fully lived up to her legendary reputation as the West's preeminent fire mountain. A long eruptive episode, from 1480 to about 1630, strikingly reformed the volcano's profile, fashioning the elegant symmetry that inspired native tales of her feminine allure. Renewed explosive outbursts and the growth of the Goat Rocks dome high on the north flank between 1800 and 1857 did not seriously mar her conical perfection. But the 1980 eruptions, which blasted away 1,300 feet from the summit and destroyed the entire north side, took away her beauty, bringing the saga of Loowit full circle.

The Klickitat imagination transformed chaotic events into roman-tic legend. Their tales showed that the histories of seemingly placid river and quiet snowpeak were punctuated by sudden and violent change. As the saga of river and mountain continues, Loowit may regain her lost beauty or reawaken her sleeping lovers, Adams and Hood. Or the Great Spirit may send another great earthquake to convulse the Pacific Northwest, perhaps creating a new Bridge of the Gods.

Now tranquil, 6,900 years ago Crater Lake was the scene of the most cataclysmic volcanic eruption to occur in the West since the end of the Ice Age. Encircling lava walls rise nearly 2,000 feet above the surface of the lake, which is 1,932 feet deep. Llao Rock, center right, is a thick lava flow named for the Indian god of the Underworld. Wizard Island, a cinder cone rising 800 feet above the lake, is said to represent the maiden who rejected Llao's advances. —Oregon State Highway Department

Chapter 29

GEOMYTHOLOGY:
The Battle of Llao and Skell

American Indians commonly interpreted natural catastrophes as the actions of invisible spirits who controlled the physical world. The Klamath Indians of southern Oregon devised a remarkable myth to explain the origin of Crater Lake. In 1865 Lalek, an aged Klamath chief, passed on this tale to William M. Colvig, a young soldier then stationed at Fort Klamath, Oregon. Narrated in the Chinook jargon, Lalek's account is extraordinary for the vast length of time it had been transmitted from generation to generation and for its grasp of geologic facts then unknown to Caucasians.

Lalek emphasized the extreme antiquity of the catastrophe he described, wryly noting that it occurred long ago when his people lived in rock houses and white men ran wild in the woods. It happened before the stars fell, when the spirits of earth, sea, and sky still spoke to human beings. In those days Llao, Chief of the Below World, would come up from his dwelling inside the earth and stand atop the high mountain that then towered over southern Oregon. One day the Chief saw Loha, a beautiful young woman, and begged her to return with him to his lodge in the Underworld. Even though he promised her eternal life, the maid refused to go with him.

When the wise men of her people's council would not force the girl to accept Llao, he thundered angrily and tried to destroy Loha's people with fire. But Skell, Chief of the Above World, heard Llao's threat and descended from heaven to the summit of Mount Shasta, which stands almost 125 miles to the south. In the conflict that followed, all the spirits of earth and sky waged war. The sky glowed with sheets of flame, then turned dark as night while the earth heaved convulsively. Torrents of fire flowed from Llao's mountain to burn the forests, reaching even the lodges of Loha's tribe, driving her people to flee into the waters of Klamath Lake.

Two brave medicine men determined to save their people from Llao's curse. Carrying torches, the old men climbed Llao's mountain and hurled themselves into the fiery mouth of the Underworld. Noting their sacrifice, the Chief of the Above World again shook the earth, causing Llao's mountain to crash down upon him. When the dark clouds of ash cleared and light returned, the lofty peak was gone, a giant hole in its place. The curse of fire was lifted and the great mountain basin gradually filled with rainwater, forming Crater Lake.

Readers familiar with Greek mythology may find in Llao's unrequited passion for Loha a striking parallel to the classical tale about Hades, god of the Underworld, who similarly pursued a reluctant

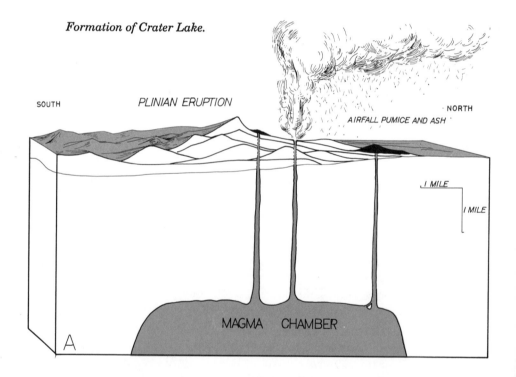

Formation of Crater Lake.

A. Stage 1: A Plinian eruption column is ejected vertically from a single vent north of Mt. Mazama's main summit. Ashfall extends hundreds of miles to the northeast.

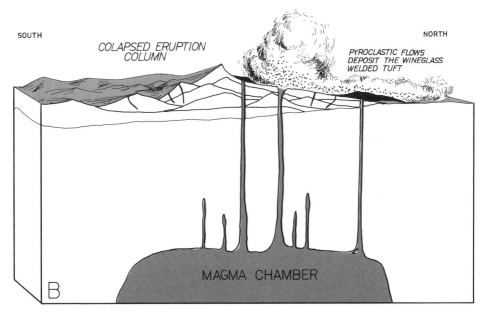

B. Stage 2: Overburdened with fragmental material, the Plinian column collapses, triggering the first major pyroclastic flow (the Wineglass Welded Tuff).

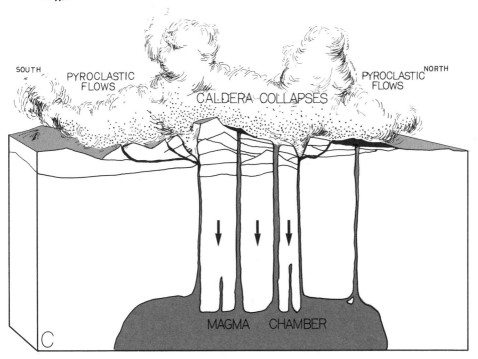

C. Stage 3: As the underground magma chamber is drained, Mt. Mazama's former summit collapses. New vents open along "ring fractures," discharging voluminous pyroclastic flows that sweep down all sides of the volcano.

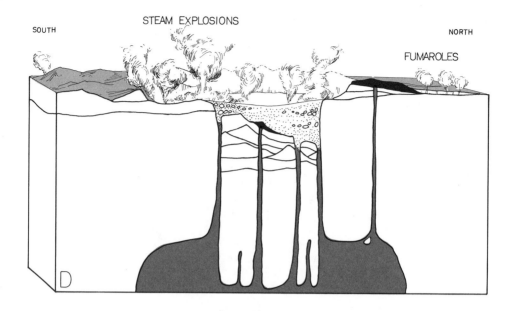

D. Stage 4: During the final phase, steam explosions deposit large quantities of ash, partly filling the caldera.

CRATER LAKE AT PRESENT LEVEL

SOUTH   NORTH

GARFIELD PK   WIZARD ISLAND   THE WATCHMAN   LLAO ROCK

LAKE LEVEL

E

E. The present: Post-caldera eruptions have built Wizard Island and other submerged cones on the caldera floor. Rain and snow melt have half-filled the basin, forming Crater Lake.

maid, Persephone, carrying her off to his subterranean kingdom. The myth's illumination of the geologic record is equally notable. In 1865, when Colvig first heard Lalek's story, Crater Lake was virtually unknown to whites and its geologic origin shrouded in mystery. Only with the work of pioneer geologist J. S. Diller (1902) and the classic study of Howel Williams (1942), did the sequence of events leading to the lake's creation become clear.

The Llao-Skell myth preserves an amazing number of geologic facts. Before about 6,900 years ago, a large volcanic cone, posthumously named Mount Mazama, rose more than a mile above the present level of Crater Lake. After having slept peacefully for 15,000 to 20,000 years, Mount Mazama suddenly produced the most violently explosive eruption to occur in North America since the Ice Age ended 10,000 years ago. Ash clouds shot twenty miles or more into the stratosphere and deposited a beige-orange tephra over most of the Pacific Northwest. Then Mazama literally boiled over. Dozens of new vents spewed flows of incandescent pumice, creating rivers of fire down all sides of the volcano. Incinerating everything in their paths, the pyroclastic flows traveled forty miles or more from their source. Some poured into Klamath Marsh, from which floating rafts of pumice eventually washed down into the Klamath Lakes. No wonder Lalek's ancestors sought refuge in the cool lake waters!

After Mazama thundered its last and the ashfall cleared, the survivors discovered that Llao's mountain had disappeared. In its place was an enormous caldera, five by six miles in diameter and up to 4,000 feet deep. This basin slowly filled with water, forming the bluest and deepest lake (1,932 feet) in the United States.

The Klamath myth correctly states that Mazama did not lose its former summit by blowing itself apart. The volcano erupted about forty cubic miles of pumice, emptying much of its underground magma reservoir. As the roof of the magma chamber subsided, it removed support from the volcanic cone, allowing it to collapse.

Llao's house crumbled, but Skell's mountain is still a pedestal to which the sky god can step down from heaven. During the last 10,000 years Mount Shasta has erupted frequently, building the large cone of Shastina on its western flank and erecting its summit dome, elevation 14,161 feet. The Klamath myth accurately reports that Skell has visited Shasta often, kindling fires even during historic time.

Chief Lalek was also right in noting that Llao's collapse lifted the curse of fire. After Mazama's summit was destroyed, volcanic activity was restricted to the interior of the caldera, where intermittent eruptions built several cones now submerged beneath the lake. The latest major activity formed Wizard Island, the beautifully symmetrical cinder cone that rises 760 feet above the lake surface. One version of the myth reports that this youthful cone represents the maid whom Llao loved.

Llao now sulks deep underground, but his passion has not wholly cooled. A 1989 survey of Crater Lake's floor reveals that heat sources raise deep water temperatures nearly thirty degrees above their normal levels. Fallen Llao and triumphant Skell are only dozing. The gods of earth and sky could reassert their destructive power at any time.

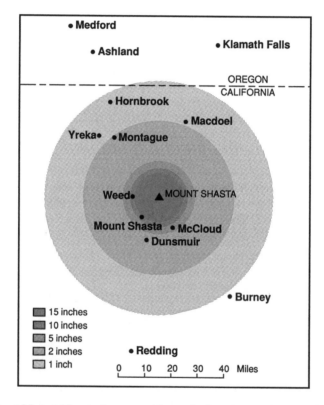

Hazards at Mount Shasta from eruptions of volcanic ash, based on the largest ash eruption of the last 10,000 years. Note that projected ash deposits thin progressively with distance from the summit. Because of prevailing westerly winds, most ash will probably fall east of the volcano.
—After Crandell, 1989

Chapter 30

COSMIC AND
GEOLOGIC VIOLENCE:
Learning To Cope With Chaos

On March 23, 1989, an asteroid half a mile in diameter zipped past Earth at 50,000 miles per hour, missing our planet by a mere 500,000 miles. If the wayward visitor had collided with Earth, astronomers estimate that the impact would have been equivalent to the explosion of 20,000 hydrogen bombs. A direct hit on land would have opened a crater five to ten miles across, large enough to obliterate an entire city.

Such cosmic close calls are probably common in Earth's long history, but a growing body of evidence suggests that Earth does not always escape the impact of solar bombshells. The periodic mass extinctions of most life forms that punctuate the geologic record appear to occur suddenly, perhaps as the result of a cosmic disaster, a large asteroid or comet striking our planet.

Astronomers estimate that about 1,500 asteroids and comets large enough to decimate the Earth are careening through the solar system. During this century several have come close to hitting home. Besides the near-miss of 1989, other objects have rushed by frighteningly close in 1932 and 1937. In 1908 an extraterrestrial object, perhaps the fragment of an old comet, devastated a large area near Stony Tunguska River in Siberia. About 50,000 years ago, a larger piece of interplanetary debris struck Arizona, creating the mile-wide Meteorite Crater.

Some space scientists recently noted that asteroids approximately two miles in diameter reach the vicinity of the earth every 300,000 years. Assuming that any one of them may have crashed, the scientists calculated that during an average human life span of fifty years, each individual has roughly one chance in 6,000 of becoming the target of an asteroid hit. According to this calculation, the average person runs a higher risk of dying by asteroid impact than of death from an airplane crash, the latter estimated to be one chance in 20,000. To put the matter in perspective, an insurance company estimates that one's chance of being killed in an automobile accident is about one in 100.

Consider the effects of a meteorite six to ten miles in diameter plunging deep into the continental crust: rock in the immediate vicinity would be vaporized, that farther away blasted into powder. Earthquakes too powerful to measure on any known seismic scale would convulse much of the globe, hurtling people and objects into the air thousands of miles from the target area. A column of fire soaring many miles into the stratosphere would incinerate structures and forests throughout the continent, perhaps igniting a worldwide conflagration.

Even if the inferno were confined to a single continent, the ensuing darkness and cold would be global. Dust and soot from the impact area would envelop the earth like a black shroud. As dust from the impact plume spread over the upper atmosphere, it would create an inch-thick canopy that would repel sunlight, drastically lowering world temperatures. After three to six months the freezing darkness would dissipate, but acid rainfall would continue to poison survivors.

The mass annihilation of ocean plankton would cause a final stage in the catastrophe, global warming. Without plankton to consume carbon dioxide, the amount of that gas in the atmosphere would greatly increase, raising average world temperatures by several degrees. Polar icecaps and other glaciers formed during the sub-zero darkness would rapidly melt and flood coastal areas. It would take many millennia for plankton to thrive again and gradually restore earth to normal temperatures.

Moderate-sized impacts seem to be randomly distributed in time, but many scientists believe that the catastrophic impacts that produce mass extinctions of plant and animal life happen at regularly spaced intervals. Recent studies indicate that large asteroids or comets hit Earth periodically, every 26 or 31 million years. The stony asteroids, which form a belt circling the Sun between Mars and Jupiter, may be fragments of a planet that never coalesced from primordial dust clouds. When collisions between these mountain-sized rocks push an asteroid out of orbit, it may be flung toward the Sun and into Earth's vicinity.

Another potential source of cosmic missiles targeting Earth lies at the outermost reaches of the solar system, far beyond the orbit of Pluto. The Oort Cloud contains numerous comets, huge dirty snow balls that can be thrown out of their orbits by the gravitational pull of passing stars. The Sun's passage through spiral arms of the galaxy may dislodge some comets from their normal paths and draw them into the inner solar system, where a few collide with Earth.

If periodic comet showers are responsible for the impacts that trigger mass extinctions, they may occur in a series of events closely spaced in time. Although earth scientists have identified relatively few impact sites, all but one on land, at least three target areas may be associated with the best-known extinction, that at the Cretaceous-Tertiary boundary of about 65 million years ago. The Deccan basaltic flood eruptions of that age may have been triggered by an impact in western India. The slightly younger Brito-Arctic eruptions, which spread lava over the northern British Isles and Greenland and generated the hot spot that now underlies Iceland, may be another impact site. A third site is near Manson, Iowa, where a 65 million year-old crater twenty-five miles in diameter lies buried beneath several hundred feet of rubble from an Ice Age glacier.

The meteorite impact theory is intensely controversial and may not be settled for years, but it is undeniable that earth has repeatedly experienced catastrophic changes over remarkably brief spans of time. Whatever their cause, the sudden changes radically restructure the global environment and hence the course of evolution. Numerous species that once dominated the earth are ruthlessly eliminated, allowing previously obscure species to proliferate and replace them. Earth's biosphere, teeming with infinitely varied life forms, is periodically wiped clean, forcing life to start over along entirely new lines. Dinosaurs and other reptiles that reigned supreme for millions of years are ruthlessly swept away and mammals become new lords of the earth, at least temporarily.

The fact that over ninety-nine percent of all plants and animals that once existed are now extinct challenges human assumptions about the regularity and stability of the world we inhabit. Certain processes of nature, such as the earth's motion in orbit around the sun and the cycle of the seasons, seem to be indefinitely repeatable and therefore comfortably predictable. Other events, such as meteorite impacts, mass extinctions, catastrophic earthquakes, ice ages and other drastic changes in global climate, do not form a regular pattern. These sudden and commonly violent intrusions of disorder into ostensibly stable systems—chaos—have repeatedly transformed the planet. Typically occurring at irregular intervals over long eons of geologic time, chaotic forces destroy or reshape ancient landforms and long-established ecological communities. Global changes may benefit a few life forms, but the majority perish utterly, a chilling precedent for the human race.

In ancient creation accounts, the ordered world, cosmos, develops out of primordial chaos, the original dark void that many early cultures saw as constantly threatening to reassert its power over nature and society. By contrast, classical philosophers and modern scientists confidently postulated a cosmic order that operates immutably according to knowable and predictable laws. New theories of chaos, however, which involve everything from individual human lives to astronomical and geological processes, call into question previously held concepts of intelligible order.

Some scientists believe that we can take steps to protect ourselves from the threat of natural violence. The world's astronomers could establish an observation network scanning the skies for asteroids and comets that might intercept the Earth's orbit. When identified, the size and projected path of the objects could be calculated, giving years of warning before they reach the planet. It may now theoretically be possible to launch piloted missions to an approaching asteroid and set off rockets that will push it sufficiently off course to bypass Earth. While some astronomers doubt the feasibility of such a plan, pointing out that an object directly aimed at Earth would not be visible until too late, establishing an anti-asteroid program may eventually become a top governmental priority. No matter how costly, it could be our only means of avoiding the fate of the dinosaurs.

However we cope with future astronomical threats, we can now effectively prepare to mitigate the effects of other chaotic events. In some parts of the United States, preparation may be as simple as not building on top of recognized earthquake fault zones. In Daly City, just south of San Francisco, irresponsible developers have constructed large housing tracts directly over the San Andreas fault, an area that will experience shaking of maximum intensity when the fault next shifts. Other developments stand at the edge of steep cliffs overlooking the Pacific Ocean, landforms likely to crumble and slide during an earthquake. The legislation and enforcement of safe building codes will not only produce earthquake-resistant structures, it will also help prevent many deaths. The collapse of the Interstate 880 Cypress structure and the extensive damage to other northern California freeways in 1989 shows that many older bridges and overpasses must either be replaced or reinforced to withstand earthquakes. The good news of '89 was that newer, well-designed highways and other structures stood up well to violent shaking, demonstrating that good construction can limit damage in future shocks. The prevalence of old, unreinforced masonry buildings throughout the country remains one of the major hazards to the population. Even in California, which has one of the nation's strictest building codes, there are an estimated 30,000 such structures, 8,000 in the quake-prone Los Angeles region alone.

Heightened public awareness of seismic hazards in a citizen's own neighborhood can also help forestall loss of life and property. Recognizing that survival is not luck, residents can significantly minimize earthquake damage to their homes and business structures. Most casualties result from partial building collapse, falling objects and debris, such as toppling bookcases, heavy picture frames, ceiling plaster, and light fixtures. Many injuries can be avoided by securing these items before the quake occurs.

In almost every earthquake thousands of private houses are badly damaged by being thrown off their foundations. The California State Office of Emergency Services points out that homeowners can save millions of dollars in losses by using a simple and cheap method to brace the base of a house against destructive shaking. As state inspectors found after the 1984 Morgan Hill quake, a moderate tremor on the Calaveras fault, near San Jose, California, much of the damage could have been prevented. As the law requires in many earthquake-prone areas, the heavy wooden "sill plates" that lie atop the concrete foundations were all properly bolted to the foundations. But the vertical studs above the sills, that support the first-floor beams and joists, were inadequately strengthened against side-to-side earthquake movement.

According to some California engineers, installation of plywood sheets by nailing them securely across the studs above the foundation near each corner of a house will keep the house from slipping off its foundation during earthquakes. Such preventive measures should cost no more than a few hundred dollars and can save a homeowner many thousands of dollars when a destructive quake strikes.

By refusing to live on fault zones, unconsolidated soils, terrain susceptible to sliding, or low-lying coastal areas subject to tsunamis, people can greatly reduce their potential earthquake losses. The same principle of prudent site selection will also mitigate the danger from volcanic eruptions. Communities located on the floors of valleys that head on a potentially active volcano are in great jeopardy. Valley floors, in some places up to several tens of miles from the volcano, are typically subject to inundation from sudden snowmelt, mudflows, and avalanches. Some volcanoes, such as Glacier Peak, Shasta, and St.

Helens, commonly erupt swift-moving pyroclastic flows that can incinerate everything in their paths for five to ten miles beyond their source. Hot ash clouds accompanying pyroclastic flows can sweep over ridgetops as well as down stream valleys. Fallout from ash plumes ejected high into the stratosphere can blanket thousands of square miles downwind from the volcano, temporarily reducing visibility in the affected areas to near zero and bringing air and ground traffic to a halt.

When a volcano, whether in the Cascades, the Mono Lake region, or elsewhere, gives signs of an impending eruption, the wisest course is to move out of the area until after the activity is safely over. The common American tendency to pursue business as usual, to regard a familiar landform as a non-threat and to ignore geologists' warnings, is likely to produce many unnecessary deaths and injuries. As in reducing the adverse effects of earthquakes, an enlightened approach to land-use planning and a far-sighted government policy for relief services after a volcanic eruption will help mitigate the toll.

Even the most disciplined and well-regulated human life is vulnerable to the sudden intrusion of chaos, whether in the form of a crippling accident, severe illness, personal bereavement, or the unexpected loss of a job or life partner. Geologic violence produces chaos on a larger scale and threatens to unravel the very fabric of society. Besides the widespread physical suffering they typically cause, chaotic events are damaging psychologically because their randomness and destructiveness do not fit into any coherent human theology or philosophy. In 1755 the Lisbon earthquake killed 50,000 people, demonstrating to Europe's believers in rational purpose that we do not live in the best of all possible worlds. Nature cares little for human welfare either individually or collectively.

Although the erratic mindlessness of chaotic events threatens cherished notions of order and purpose, society as a whole typically shows great resilience in recovering from natural disasters. Despite high tolls in individual losses, cities ravaged by earthquake, fire, or flood are quickly rebuilt. Within three years of the 1906 holocaust San Franciscans had constructed a new city. Unfortunately society's collective response to catastrophe usually is to forget that it ever happened as quickly as possible and to ignore the possibility of its recurrence. Optimism born of amnesia is a dangerous luxury, however. Eschewing either complacency or panic, Americans can profit from the lessons taught by St. Helens and the many damaging earthquakes that strike different parts of the United States. A great earthquake, much stronger than the Loma Prieta shock of 1989, has not devastated a large region of the country since 1906, but will inevitably occur—in California, Alaska, the Pacific Northwest, the central Mississippi Valley, or the eastern states. As violent and uncaring as a barbarian invader, chaos eventually will smash into nearly all our neighborhoods. Whether it brings fire from heaven, or convulses the earth beneath our feet, we will never again be as we were before.

Bibliography

Chapter 1. Living with Chaos

Blong, R. J. 1984. *Volcanic Hazards: A Sourcebook on the Effects of Eruptions.* Sydney, Australia: Academic Press.

Burke, J. G. 1986. *Cosmic Debris: Meteorites in History.* Berkeley and Los Angeles: University of California Press.

Crandell, D. R.; Booth, B.; Kusumadinata, K.; Shimozuru, D.; Walker, G. P. L.; and Westercamp, D. 1984. *Sourcebook for Volcanic Hazards Zonation.* Paris: Unesco.

Lunar and Planetary Institute. 1988. *Global Catastrophes in Earth History: An Interdisciplinary Conference on Impacts, Volcanism, and Mass Mortality, Snowbird, Utah.* Lunar and Planetary Institute and the National Academy of Sciences.

Nance, J. J. 1988. *On Shaky Ground: An Invitation to Disaster*, New York: William Morrow and Company, Inc.

Steinbrugge, Karl V. 1982. *Earthquakes, Volcanoes, and Tsunamis: An Anatomy of Hazards.* New York: Skandia America Group.

Chapter 2. Experiencing An Earthquake

Bolt, Bruce A. 1987. *Earthquakes.* rev. ed. Berkeley, University of California Press.

Carder, D. S., ed. 1965. *Earthquake Investigations in the Western United States, 1934-1964.* Washington, D.C.: U.S. Coast and Geodetic Survey, U.S. Government Printing Office.

Edwards, H. H. 1951. Lessons in Structural Safety Learned from the 1949 Northwest Earthquake. *Western Construction* 26(2): 70-74; no. 3, 85-88; no. 4, 90-92.

Harris, Stephen L. 1985. Earthquake Hazards in the West. *American West,* 22 (3): 28-36.

Murphy, L. M., and Ulrich, F. P. 1951. *United States Earthquakes 1949.* U.S. Coast and Geodetic Survey, Serial no. 748. Washington, D.C.: U.S. Government Printing Office.

Nuttli, O.W. 1952. The Western Washington Earthquake of April 13, 1949. *Bulletin of the Seismological Society of America.* 42: 21-28

Schnell, Mary L., and Herd, D. G. 1983. *National Earthquake Hazards Reduction Program: Report to the United States Congress.* United States Geological Survey Circular 918.

Spence, W.; Simpkin, S. A.; and Choy, G. L. 1989. Measuring the Size of an Earthquake. *Earthquakes and Volcanoes* 21 (1): 58-63.

Steinbrugge, Karl V. 1982. Earthquake Experience in the United States and Canada. In *Earthquakes, Volcanoes, and Tsunamis: An Anatomy of Hazards*, Steinbrugge, ed. New York: Skandia America Group.

Stover, C. W. 1989. Evaluating the Intensity of United States Earthquakes. *Earthquakes and Volcanoes* 21(1): 45-53.

U.S. Coast and Geodetic Survey. 1965. *The Puget Sound, Washington, Earthquake of April 29, 1965.* Washington, D.C.: U.S. Government Printing Office.

von Hake, Carl A. 1978. Earthquake History of Washington. *Earthquake Information Bulletin* 10 (1): 28-34.

Chapter 3.   Plates in Motion: Our Dynamic Earth

Greenwood, P. H., and Cocks, L. R. M. 1981. *The Evolving Earth.* Cambridge: Cambridge University Press.

McAlester, A. L.; Eicher, D. L.; and Rottman, M. L. 1984. *The History of the Earth's Crust.* Englewood Cliffs, N. J.: Prentice-Hall.

Plummer, C. C., and McGeary, D. 1985. *Physical Geology.* 3rd ed. Dubuque, Iowa: Wm. C. Brown Publishers.

Chapter 4.   The West Is a Geologic Crazy Quilt

Alt, D. D., and Hyndman, D. W. 1984. *Roadside Geology of Washington.* Missoula, Montana: Mountain Press Publishing Company.

Dickinson, W. R. 1981. Plate Tectonics and the Continental Margin of California. In *The Geotectonic Development of California*, ed. W. G. Ernst. Englewood Cliffs, N.J.: Prentice-Hall.

Jones, D. L.; Cox, A.; Coney, P.; and Beck, M. 1980. The Growth of Western North America. *Scientific American*, Nov. 1980.

McAlester, A. L.; Eicher, D. L.; and Rottman, M.L. 1984. *The History of the Earth's Crust.* Englewood Cliffs, N.J.: Prentice-Hall.

McFee, J. A. 1983. *In Suspect Terrain.* New York: Farrar, Straus & Giroux.

Redfern, Ron. 1983. *The Making of a Continent.* New York: Times Books

Sullivan, Walter. 1985. Pieces of a Global Jigsaw Puzzle. *Smithsonian* 15 (10): 66-74.

Chapter 5.   The 1989 California Earthquake

McNally, Karen C., et al. 1989. Santa Cruz Mountains (Loma Prieta) Earthquake. *Eos*, Nov. 7, 1989, 1466-67.

McNutt, Steve. 1990. Loma Prieta Earthquake, October 17, 1989, Santa Cruz County, California. *California Geology* 43 (1): 3-7.

Montgomery, David R. 1990. Effects of Loma Prieta Earthquake, October 17, 1989. *California Geology* 43 (1): 8-13, 24.

Plafker, George, and Galloway, J. P., eds. 1989. *Lessons Learned from the Loma Prieta Earthquake of October 17, 1989.* U.S. Geological Survey Circular 1045.

Ramseyer, Cynthia. 1989. *The Loma Prieta Earthquake.* U.S. Geological Survey Open-File Report 89-687. Includes 36 slides.

San Francisco Chronicle. October 18, 1989, through December 31, 1989.

San Francisco Chronicle. 1989. *The Quake of '89.* San Francisco: Chronicle Books.

Chapter 6. California's Deadly Earthquakes

Agnew, D. C., and Sieh, K. E. 1978. A Documentary Study of the Felt Effects of the Great California Earthquakes of 1857: *Bulletin of the Seismological Society of America* 68 (6): 1717-29.

Gilbert, G. K.; Humphrey, R. L.; Sewell, J. S.; and Soule, F. 1907. *The San Francisco Earthquake and Fire of April 18, 1906, and their Effects on Structures and Structural Materials.* U.S. Geological Survey Bulletin 324.

Hansen, Gladys, and Condon, Emmet. 1989. *Denial of Disaster.* San Francisco: Cameron and Company.

Kennedy, John Castillo. 1963. *The Great Earthquake and Fire, San Francisco, 1906.* New York: William Morrow and Company.

Lawson, A. C., et al. 1908. *The California Earthquake of April 18, 1906, Report of the State Earthquake Commission, 2 vols. and atlas.* Washington, D.C.: Carnegie Institute of Washington.

National Board of Fire Underwriters. 1906. *The San Francisco Conflagration of April, 1906.* Special Report to the National Board of Fire Underwriters.

Oakeshott, G. B., ed. 1955. *Earthquakes in Kern County, California, During 1952.* California Division of Mines and Geology, Bulletin 171.

Saul, Eric, and Denevi, Don. 1981. *The Great San Francisco Earthquake and Fire, 1906.* Millbrae, California: Celestial Arts.

Thomas, Gordon, and Witts, Max M. 1971. *The San Francisco Earthquake.* New York: Stein and Day.

United States Geological Survey and the National Oceanic and Atmospheric Administration. 1971. *The San Fernando, California, Earthquake of February 9, 1971.* United States Geological Survey Professional Paper 733.

Wood, H. O. 1933. Preliminary Report on the Long Beach Earthquake, *Bulletin of the Seismological Society of America* 23: 43-56.

Chapter 7. When The "Big One" Hits California

Davis, J. F.; Bennett, M. H.; Borchardt, G. A.; Kahle, J. E.; Rice, S. J.; and Silva, M. A. 1982. *Earthquake Planning Scenario for a Magnitude 8.3 Earthquake on the San Andreas Fault in Southern California*. California Division of Mines and Geology Special Publication 60.

Davis, J. F.; Bennett, J. H.; Borchardt, G. A.; Kahle, J. E.; Rice, S. J.; and Silva, M. A. 1982. *Earthquake Planning Scenario for a Magnitude 8.3 Earthquake on the San Andreas Fault in the San Francisco Bay Area*. Special Publication 61, California Division of Mines and Geology.

Evernden, J. F.; Fumal, T. E.; Harp, E. L.; Hatzell, S. H.; Joyner, W. B.; Keefer, D. K.; Spudich, P. A.; Tinsley, J. C.; Yerkes, R. F.; and Youd, T. L. 1985. *Predicted Geologic and Seismologic Effects of a Postulated Magnitude 6.5 Earthquake along the Northern Part of the Newport-Inglewood Zone*. In: Ziony, J. I., ed. 1985. *Evaluating Earthquake Hazards in the Los Angeles Region—An Earth Science Perspective*. U.S. Geological Professional Paper 1360, 415-42.

Hansen, Gladys, and Condon, Emmet. 1989. *Denial of Disaster*, San Francisco: Cameron and Company.

Heppenheimer, T. A. 1988. *The Coming Quake: Science and Trembling on the California Earthquake Frontier*. New York: Times Books.

Holden, R., Lee, R., and Reichle, M. 1989. *Technical and Economic Feasibility of an Earthquake Warning System in California*. California Division of Mines and Geology. Special Publication 101.

Nance, J. J. 1988. *On Shaky Ground: An Invitation to Disaster*. New York: William Morrow and Company, Inc.

Scawthorn, Charles. 1987. *Fire Following Earthquake: Estimates of the Conflagration Risk to Insured Property in Greater Los Angeles and San Francisco*. Oak Brook, Illinois: All-Industry Research Council.

Steinbrugge, K. V.; Bennett, J. H.; Lagorio, H. J.; Davis, J. F.; Borchardt, G.; and Toppozada, T. R. 1987. *Earthquake Planning Scenario for a Magnitude 7.5 Earthquake on the Hayward Fault in the San Francisco Bay Area*. California Division of Mines and Geology, Special Publication 78.

Toppozada, Tousson R.; Bennett, J. H.; Borchardt, Glenn; Saul, Richard; and Davis, J. F. 1989. Earthquake Planning Scenario for a Major Earthquake on the Newport-Inglewood Fault Zone. *California Geology* 42 (4): 75-84.

Turner, R. H.; Nigg, J. M.; and Paz, D. H. 1986. *Waiting for Disaster: Earthquake Watch in California*. Berkeley and Los Angeles: University of California Press.

Ziony, J. I., ed. 1985. *Evaluating Earthquake Hazards in the Los Angeles Region—An Earth Science Perspective*. United States Geological Survey Professional Paper 1360.

Chapter 8. Plates in Motion: Alaska's Superquakes

National Academy of Sciences, Division of Earth Sciences, Committee on the Alaska Earthquake. 1968-1973. *The Great Alaska Earthquake of 1964.* Vol. 1, *Geology.* Vol. 2, *Seismology and Geodesy.* Vol. 3, *Hydrology.* Vol. 4, *Biology.* Vol. 5, *Oceanography and Coastal Engineering.* Vol. 6, *Engineering.* Vol. 7, *Human Ecology.* Vol. 8, *Summary and Recommendations.* Washington, D.C.: National Academy of Sciences.

Schnell, Mary L., and Herd, D. G. 1983. *National Earthquake Hazards Reduction Program: Report to the United States Congress.* U.S. Geological Survey Circular 918.

Steinbrugge, Karl V. 1982. *Earthquakes, Volcanoes, and Tsunamis: An Anatomy of Hazards.* New York: Skandia America Group.

U.S. Coast and Geodetic Survey, Environmental Sciences Services Administration. 1967. *The Prince William Sound, Alaska, Earthquake of 1964 and Aftershocks.* Vol. 1, *Operational Phases of the USC & GS, Including Seismicity.* 1966. Vol. 2, Part A: *Engineering Seismology.* 1967. Vol. 2, Part B: *Seismology.* and Vol. 3, *Geodesy and Photogrammetry Research Studies,* Washington, D.C.: U.S. Government Printing Office.

U.S. Geological Survey. 1964-1968. *The Alaska Earthquake, March 27, 1964.* U.S. Geological Survey Professional Papers 541, 542, 543, 544, and 545.

U.S. Geological Survey. 1964. *Alaska's Good Friday Earthquake, March 27, 2964.* Geological Survey Circular 491.

Walker, Bryce. 1982. *Earthquake.* Alexandria, Virginia: Time-Life Books.

Chapter 9. Giant Earthquakes:
A Possibility for the Pacific Northwest

Atwater, Brian F. 1987. Evidence for Great Holocene Earthquakes Along the Outer Coast of Washington State. *Science* 236 (4804):`942-944.

Hays, Walter W. 1988. *Workshop on "Evaluation of Earthquake Hazards and Risk in the Puget Sound and Portland Areas," Olympia, Washington, April 12-15, 1988, Proceedings of Conference XLII.* U.S. Geological Survey Open-File Report 88-541.

Heaton, Thomas H, and Hartzell, S. H. 1987. Earthquake Hazards on the Cascadia Subduction Zone. *Science* 236: 162-68.

Heaton, Thomas H., and Snavely, P. D., Jr. 1985. Possible Tsunami Along the Northwestern Coast of the United States Inferred from Indian Traditions. *Bulletin of the Seismological Society of America* 75 (5): 1455-60.

Third Annual Puget Sound/Portland Area Workshop on Earthquake Hazard and Risk. March 28-30, 1989. *Abstracts with Programs.* Oregon Department of Geology and Mineral Industries, U.S. Geological Survey.

von Hake, Carl A. 1978. Earthquake History of Washington. *Earthquake Information Bulletin* 10 (1): 28-34.

Chapter 10.   The Earth Stretches Its Skin:
Death Valley and The Great Basin

Fiero, Bill. 1986. *Geology of the Great Basin*. Reno, Nevada: University of Nevada Press.

Hildreth, Wes. 1976. *Death Valley Geology: Rocks and Faults*. Death Valley, California: The Death Valley Natural History Association.

Chapter 11.   Earthquakes in Unexpected Places:
The Catastrophic New Madrid Quakes of 1811-1812

Dutton, C. E. 1887-1888. *The Charleston Earthquake of August 31, 1886*. U.S. Geological Survey, Ninth Annual Report.

Fuller, M. L. 1912. *The New Madrid Earthquake*. U.S. Geological Survey Bulletin 494.

McKeown, F. A., and Pakiser, L. C., eds. 1982. *Investigations of the New Madrid, Missouri, Earthquake Region*. U.S. Geological Survey Professional Paper 1236.

Nuttli, Otto W. 1973. The Mississippi Valley Earthquakes of 1811 and 1812: Intensities, Ground Motion and Magnitudes. *Bulletin of the Seismological Society of America* 63: 227-48.

Nuttli, Otto W.; Bollinger, G.A.; and Herrmann, R.B. 1986. *The 1886 Charleston, South Carolina, Earthquake: A 1986 Perspective*. U.S. Geological Survey Circular 985.

Penick, J. L. 1981. *The New Madrid Earthquakes of 1811-1812*. Columbia, Missouri: The University of Missouri Press.

Chapter 12. Where and Why Destructive Earthquakes
Will Strike Next

Algermissen, S. T., and Perkins, D. M. 1976. *Probabilistic Estimate of Maximum Acceleration in Rock in the Coterminous United States*. U.S. Geological Survey Open-File Report 76-416.

San Francisco Chronicle. Oct. 5, 1988. Big Quake Likely to Hit East by the Year 2,000.

Davis, J. F.; Bennett, M. H.; Borchardt, G. A.; Kahle, J. E.; Rice, S. J.; and Silva, M. A. 1982. *Earthquake Planning Scenario for a Magnitude 8.3 Earthquake on the San Andreas Fault in Southern California*. California Division of Mines and Geology, Special Publication 60.

Davis, J. F.; Bennett, J. H.; Borchardt, G. A.; Kahle, J. E.; Rice, S. J.; and Silva, M. A. 1982. *Earthquake Planning Scenario for a Magnitude 8.3 Earthquake on the San Andreas Fault in the San Francisco Bay Area*. California Division of Mines and Geology, Special Publication 61.

Evernden, J. F.; Fumal, T. E.; Harp, E. L.; Hatzell, S. H.; Joyner, W. B.; Keefer, D. K.; Spudich, P. A.; Tinsley, J. C.; Yerkes, R. F.; and Youd, T. L. 1985. *Predicted Geologic and Seismologic Effects of a Postulate Magnitude 6.5 Earthquake Along the Northern Part of the Newport-Inglewood Zone.* In Ziony, J. I., ed. *Evaluating Earthquake Hazards in the Los Angeles Region—An Earth Science Perspective.* U.S. Geological Survey Professional Paper 1360, 415-42.

Hays, Walter W., ed. 1988. *Workshop on "Evaluation of Earthquake Hazards and Risk in the Puget Sound and Portland Areas," Olympia, Washington April 12-15, 1988, Proceedings of Conference XLII.* U.S. Geological Survey Open-File Report 88-541.

Heppenheimer, T. A. 1988. *The Coming Quake: Science and Trembling on the California Earthquake Frontier.* New York: Times Books.

Jaffe, Martin; Butler, JoAnn; Thurow, Charles. 1981. *Reducing Earthquake Risks: A Planner's Guide.* American Planning Association Planning Advisory Service, Report no. 364.

Nance, J. J. 1988. *On Shaky Ground: An Invitation to Disaster.* New York: William Morrow and Company, Inc.

Steinbrugge, Karl V. 1982. *Earthquakes, Volcanoes, and Tsunamis: An Anatomy of Hazards.* New York: Skandia America Group.

Steinbrugge, K. V.; Bennett, J. H.; Lagorio, H. J.; Davis, J. F.; Borchardt, G.; and Toppozada, T. R. 1987. *Earthquake Planning Scenario for a Magnitude 7.5 Earthquake on the Hayward Fault in the San Francisco Bay Area.* California Division of Mines and Geology, Special Publication 78.

Toppozada, Tousson R.; Bennett, J. H.; Borchardt, Glenn; Saul, Richard; and Davis, J. F. 1989. Earthquake Planning Scenario for a Major Earthquake on the Newport-Inglewood Fault Zone. *California Geology* 42 (4): 75-84.

Turner, R. H.; Nigg, J. M.; and Paz, D. H. 1986. *Waiting for Disaster: Earthquake Watch in California.* Berkeley and Los Angeles: University of California Press.

Ziony, J. I., ed. 1985 *Evaluating Earthquake Hazards in the Los Angeles Region—An Earth Science Perspective.* United States Geological Survey Professional Paper 1360.

Chapter 13.   Mount St. Helens: A Lethal Beauty

Crandell, D. R. 1987. *Deposits of Pre-1980 Pyroclastic Flows and Lahars from Mount St. Helens Volcano, Washington.* U.S. Geological Survey Professional Paper 1444.

Crandell, D. R., and Mullineaux, D. R. 1978. *Potential Hazards from Future Eruptions of Mount St. Helens Volcano, Washington.* U.S. Geological Survey Bulletin 1383-C.

Decker, Robert, and Decker, Barbara. 1981. The Eruptions of Mount St. Helens. *Scientific American* 244 (3): 68-80.

Foxworthy, B. L., and Hill, Mary. *Volcanic Eruptions of 1980 at Mount St. Helens: The First 100 Days.* U.S. Geological Survey Professional Paper 1249.

Harris, Stephen L. 1988. *Fire Mountains of the West: The Cascade and Mono Lake Volcanoes*. Missoula, Montana: Mountain Press Publishing Co.

Holmes, Kenneth L. 1980. *Mount St. Helens, Lady with a Past*. Salem, Oregon: Salem Press.

Keller, S. A. C., ed. 1986. *Mount St. Helens: Five Years Later*. Cheney, Washington: Eastern Washington State University Press.

Lipman, Peter W., and Mullineaux, D. R., eds. *The 1980 Eruptions of Mount St. Helens*. U.S. Geological Survey Professional Paper 1250.

Majors, Harry M. 1980. Mount St. Helens Series. *Northwest Discovery* 1 (1 & 2).

Moore, James G., and Rice, Carl J. 1984. Chronology and Character of the May 18, 1980, Explosive Eruptions of Mount St. Helens. In *Explosive Volcanism: Inception, Evolution, and Hazards 133-142*. Geophysics Study Committee, et al. Washington, D.C.: National Academy Press.

Tilling, Robert. 1982. *Eruptions of Mount St. Helens, Past, Present, and Future*. U.S. Geological Survey Series of General Interest Publications.

Williams, Chuck. 1988. *Mount St. Helens National Volcanic Monument, a Pocket Guide for Hikers, Viewers, and Skiers*. Seattle: The Mountaineers.

Yamaguchi, David. 1985. Tree-ring Evidence for a Two-Year Interval Between Recent Prehistoric Explosive Eruptions of Mount St. Helens. *Geology* 13 (8): 354-357.

Chapter 14.   The "Other" Cascade Volcanoes

Bacon, C.R. 1983. Eruptive History of Mount Mazama and Crater Lake Caldera, Cascade Range, U.S.A. *Journal of Volcanology and Geothermal Research* 18: 57-115.

Bacon, C. R.. 1987. Mount Mazama and Crater Lake Caldera, Oregon. *Geological Society of America Centennial Field Guide—Cordilleran Section, 1987*, 301-306.

Béget, James E. 1982. Recent Volcanic Activity at Glacier Peak. *Science* 215: 1389-1390.

Cameron, K. A., and Pringle, P. T. 1987. A Detailed Chronology of the Most Recent Major Eruptive Period at Mount Hood, Oregon. *Geological Society of America Bulletin* 99: 845-851.

Crandell, D. R. 1980. *Recent Eruptive History of Mount Hood, Oregon, and Potential Hazards from Future Eruptions*. U.S. Geological Survey Bulletin 1492.

Crandell, D. R. 1983. *The Geologic Story of Mount Rainier*. U.S. Geological Survey Bulletin 1292.

Crandell, D. R.; Mullineaux, D. R.; and Miller, C. D. 1979. Volcanic Hazard Studies in the Cascade Range of the Western United States. In *Volcanic Activity and Human Ecology*. ed. Sheets, P. D., and Grayson, D. K., 195-219. New York: Academic Press.

Day, A. L., and Allen, E. T. 1925. *The Volcanic Activity and Hot Springs of Lassen Peak*. Carnegie Institute of Washington, Publication 360.

Decker, Robert, and Decker, Barbara. 1981. *Volcanoes*. San Francisco: W. H. Freeman & Company.

Driedger, C. L., and Kennard, P. M. 1986. *Ice Volumes on Cascade Volcanoes: Mount Rainier, Mount Hood, Three Sisters, and Mount Shasta*. U.S. Geological Survey Professional Paper 365.

Eppler, D. B. 1987. The May 1915 Eruptions of Lassen Peak, II: May 22 Volcanic Blast Effects, Sedimentology and Stratigraphy of Blast and Lahar Deposits, and Characteristic of the Blast Cloud. *Journal of Volcanology and Geothermal Research* 31: 65-85.

Harris, Stephen L. 1988. *Fire Mountains of the West: The Cascade and Mono Lake Volcanoes*. Missoula, Montana: Mountain Press Publishing Co.

Harris, Stephen L. 1983. In the Shadow of the Mountains. *Pacific Northwest* 17 (1): 24-33.

Harris, Stephen L. 1986. The OTHER Cascade Volcanoes: Historic Eruptions at Mount St. Helens' Sister Peaks. In *Mount St. Helens: Five Years Later*. ed. Keller, S. A. C., 20-33. Cheney, Washington: Eastern Washington State University Press.

Harris, Stephen L. 1985. Volcanic Hazards in the West. *American West* 20 (6): 30-39.

Hyde, J. H., and Crandell, D. R. 1978. *Postglacial Volcanic Deposits on Mount Baker, Washington, and Potential Hazards from Future Eruptions*. U.S. Geological Survey Professional Paper 1022-C.

Johnston, D. A., and Donnelly-Nolan, Julie. 1981. *Guides to Some Volcanic Terranes in Washington, Idaho, Oregon, and Northern California*. U.S. Geological Survey Circular 838.

Kiver, E. P. 1982. The Cascade Volcanoes: Comparisons of Geological and Historic Record. In *Mount St. Helens: One Year Later*. ed. Keller, S. A. C., 3-12. Cheney, Washington: Eastern Washington State University Press.

Loomis, B. F. 1926. *Pictorial History of the Lassen Volcano, Mineral, California*. Loomis Museum Association.

Majors, H. M. 1978. *Mount Baker: A Chronicle of Its Historic Eruptions and First Ascent*. Seattle: Northwest Press.

Miller, C. Dan. 1980. *Potential Hazards from Future Eruptions in the Vicinity of Mount Shasta Volcano, Northern California*. U.S. Geological Survey Bulletin 1503.

Williams, Howel. 1932. *Geology of the Lassen Volcanic National Park, California*. University of California Publications in Geological Science 21 (8): 195-385.

Williams, Howel. 1942. *The Geology of Crater Lake National Park, Oregon*. Carnegie Institute Publication 540.

Chapter 15. Scenery and Solitude: Lassen Volcanic National Park

Crandell, D. R.; Mullineaux, D. R.; Sigafoos, R. S.; and Rubin, Meyer. 1974. Chaos Crags Eruptions and Rockfall-Avalanches, Lassen Volcanic National Park, California. *Journal of Research U.S. Geological Survey* 2 (1): 49-59.

Day, A. L., and Allen, E. T. 1925. *The Volcanic Activity and Hot Springs of Lassen Peak*. Carnegie Institution of Washington Publication 360.

Eppler, D. B. 1987. The May 1915 Eruptions of Lassen Peak, II: May 22 Volcanic Blast Effects, Sedimentology and Stratigraphy of Blast and Lahar Deposits, and Characteristics of the Blast Cloud. *Journal of Volcanology and Geothermal Research* 31: 65-85.

Eppler, D. B., and Malin, M. C. 1989. The May 1915 Eruptions of Lassen Peak, California, I: Characteristics of Events Occurring on 19 May. In: *Proceedings in Volcanology, Volcanic Hazards*, ed. Latter, J. H. Berlin and Heidelberg: Springer-Verlag.

Finch, R. H. 1937. A Tree Ring Calendar for Dating Volcanic Events, Cinder Cone, Lassen National Park, California. *American Journal of Science* 33: 140-46.

Harris, Stephen L. *Fire Mountains of the West: The Cascade and Mono Lake Volcanoes*. Missoula, Montana: Mountain Press Publishing Company.

Loomis, B. F. 1926. *Pictorial History of the Lassen Volcano*. Mineral, California: Loomis Museum Association.

Richard, Ellis. 1988. *Lassen Volcanic: The Story Behind the Scenery*. Las Vegas, Nevada: K.C. Publications.

Chapter 16. What's Brewing Near Mono Lake?

Bailey, Roy A. 1982. Other Potential Eruption Centers in California: Long Valley-Mono Lake, Cosco, and Clear Lake Volcanic Fields. In *Status of Volcanic Prediction and Emergency Response Capabilities in Volcanic Hazard Zones of California*, ed. Martin, R. C., and Davis, J. F. California Division of Mines and Geology, Special Publication 63: 17-28.

Mader, George, and Blair, Martha. 1987. *Living with a Volcanic Threat, Portola Valley, California*. William Spangle and Associates.

Miller, C. Dan. 1985. Holocene Eruptions at the Inyo Volcanic Chain, California: Implications for Possible Eruptions in Long Valley Caldera. *Geology* 13: 14-17.

Miller, C. Dan; Mullineaux, D. R.; Crandell, D. R.; and Bailey, R. A. 1982. *Potential Hazards from Future Volcanic Eruptions in the Long Valley-Mono Lake Area, East-Central California and Southwest Nevada—A Preliminary Assessment*. U.S. Geological Survey Circular 877.

Newhall, C. G., and Dzurisin, D. 1988. *Historical Unrest at Large Calderas of the World*. U.S. Geological Survey Bulletin 1855, vol. 2.

Savage, J. C.; Cockerham, R. S.; Estrem, J. E.; and Moore, L. R. 1987. Deformation near the Long Valley Caldera, Eastern California, 1982-1986. *Journal of Geophysical Research* 92 (B3): 2721-46.

Sieh, Kerry, and Bursik, Kerry. 1986. The Most Recent Eruption of the Mono Craters, Eastern Central California. *Journal of Geophysical Research* 91 (B12): 12,539-71.

Chapter 17.   The Columbia River Plateau: Chaos From Heaven

Alt, David; Sears, J. M.; and Hyndman, D. W. 1988. Terrestrial Maria: The Origins of Large Basalt Plateaus, Hotspot Tracks and Spreading Ridges. *Journal of Geology* 96: 647-62.

Alvarez, L. W.; Alvarez, W.; Asaro, F.; and Michel, H. V. 1980. Extra-Terrestrial Cause for the Cretaceous-Tertiary Extinction. *Science* 208: 1095-1108.

Alvarez, L. W.; Alvarez, W.; Asaro, F.; and Michel, H. V. 1982. Current Status of the Impact Theory for the Terminal Cretaceous Extinction. In *Geological Interpretations of Impacts of Large Asteroids and Coments on the Earth.* ed. Silver, L. T., and Schulz, P., 305-15. Geological Society of America Special Paper 190.

Bullard, Fred. 1984. *Volcanoes of the Earth.* 2nd. rev. ed. Austin, Texas: University of Texas Press.

Lunar and Planetary Institute. 1988. *Global Catastrophes in Earth History: An Interdisciplinary Conference on Impacts, Volcanism, and Mass Mortality, Snowbird, Utah.* Lunar and Planetary Institute and the National Academy of Sciences.

Swanson, D. A., and Wright, T. L. 1981. Guide to Geologic Field Trip Between Lewiston, Idaho, and Kimberly, Oregon, Emphasizing the Columbia River Basalt Group. In *Guides to Some Volcanic Terranes in Washington, Idaho, Oregon, and Northern California,* ed. Johnston, David, and Donnelly-Nolan, Julie M. 1-28. U.S. Geological Survey Circular 838.

Chapter 18.   The Snake River Plain: Tracking A Geologic Hot Spot

Alt, David, and Hyndman, D. W. 1989. *Roadside Geology of Idaho.* Missoula, Montana: Mountain Press Publishing Company.

Bullard, Fred. 1984. *Volcanoes of the Earth.* 2nd. rev. ed. Austin, Texas: University of Texas Press: 320-30.

Kuntz, M. A.; Spiker, E. C.; Rubin, M.; Champion, D. E.; and Lefebvre, R. H. 1986. Radiocarbon Studies of Latest Pleistocene and Holocene Lava Flows on the Snake River Plain, Idaho: Data, Lessons, Interpretations. *Quaternary Research* 25: 163-76.

Kuntz, M. S.; Champion, D. E.; Spiker, E. C.; Lefebvre; R. H.; and McBroome, L. A. 1982. The Great Rift and Evolution of the Craters of the Moon Lava Field. In *Cenozoic Geology of Idaho,* ed. Bonnichsen, B., and Breckenridge, R. M. 423-32. Idaho Bureau of Mines and Geology Bulletin 26.

Smith, R. L., and Luedke, R. C. 1984. Potentially Active Volcanic Lineaments and Loci in Western Coterminous United States. In: *Explosive Volcanism: Inception, Evolution, and Hazards.* ed. Boyd, F. R., 47-66. Washington, D.C.: National Research Council Studies in Geophysics, National Academy Press.

Chapter 19. Torrents of Fire: Eruptions That Can Devastate A Continent

Alt, David, and Hyndman, D. W. 1989. *Roadside Geology of Idaho*. Missoula, Montana: Mountain Press Publishing Company.

Christiansen, R. L. 1984. *Yellowstone Magmatic Evolution: Its Bearing on Understanding Large-volume Explosive Volcanism*. In *Explosive Volcanism: Inception, Evolution, and Hazards*. ed. Boyd, F. R., 84-95. Washington, D.C.: National Research Council Studies in Geophysics, National Academy Press.

Christiansen, R. L. 1988. *The Quaternary and Pliocene Yellowstone Plateau Volcanic Field of Wyoming, Idaho, and Montana*. U.S. Geological Survey Professional Paper 729.

Fritz, William J. 1985 *Roadside Geology of the Yellowstone Country*. Missoula, Montana: Mountain Press Publishing Company.

Chapter 20. The Fiery Realm of Madame Pele: Hawaii's Volcanoes National Park

Decker, Robert, and Decker, Barbara. 1980. *Volcano Watching*. Hawaii Volcanoes National Park: Hawaii Natural History Association.

Macdonald, Gordon A., and Abbott, A. T. 1970. *Volcanoes in the Sea: The Geology of Hawaii*. Honolulu: Hawaii University Press.

Macdonald, Gordon A., and Hubbard, D. 1978. *Volcanoes of the National Parks of Hawaii*. Hawaii: Hawaii Natural History Association.

Chapter 21. The Hawaiian Hot Spot

Burke, K. C., and Wilson, J. Tuzo. 1982. Hot Spots on the Earth's Surface, In *Volcanoes and the Earth's Interior*, ed. Decker, R., and Decker, B., 31-42. Readings from Scientific American, San Francisco: W.H. Freeman and Company, p. 31-42.

Decker, Robert; Wright, T. L.; and Stauffer, P. H. eds., 1987. *Volcanism in Hawaii*. U.S. Geological Professional Paper 1350, 2 vols. Washington, D.C.: U.S. Government Printing Office.

Mullineaux, D. R.; Peterson, D. W.; and Crandell, D. R. 1987. Volcanic Hazards in the Hawaiian Islands. In *Volcanism in Hawaii*, ed. Decker, Robert, Wright, T. L., and Stauffer, P. H. 599-621. U.S. Geological Survey Professional Paper 1350, vol. 2.

Wolfe, Edward W., ed. 1988. *The Puu Oo Eruption of Kilauea Volcano, Hawaii, January 3, 1983, through June 8, 1984*. U.S. Geological Survey Professional Paper 1463. Washington, D.C.: U.S. Government Printing Office.

Chapter 22. Alaska's Mountains of Ice and Fire

Alaska Geographic. 1976. *Alaska's Volcanoes: Northern Link in the Ring of Fire* 4 (1). Edmonds, Washington: Alaska Northwest Publishing Co.

Coats, Robert R. 1950. *Volcanic Activity in the Aleutian Arc*. U.S. Geological Survey Bulletin 974-B: 35-49.

Connor, Cathy, and O'Haire, Daniel. 1988. *Roadside Geology of Alaska.* Missoula, Montana: Mountain Press Publishing Co.

Kienle, J., and Forbes, R. B. 1976. Augustine: Evolution of a Volcano, Fairbanks: *Geophysical Institute, University of Alaska, Annual Report, 1975/76* :26-48.

Kienle, J., and Swanson, S. E. 1983. Volcanism in the Eastern Aleutian Arc: Late Quaternary and Holocene Centers, Tectonic Setting and Petrology. *Journal of Volcanology and Geothermal Research* 17 :393-432.

McNutt, S. R. 1987. Eruption Characteristics and Cycle at Pavlof Volcano, Eastern Aleutians, and their Relation to Regional Earthquake Activity. *Journal of Volcanology and Geothermal Research* 31: 239-67.

Miller, T. P., and Smith, R. L. 1977. Spectacular Mobility of Ash Flows Around Aniakchak and Fisher Calderas Alaska. *Geology* 5: 173-76.

Miller, T. P., and Smith, R. W. 1987. Later Quaternary Caldera-Forming Eruptions in the Eastern Aleutian Arc, Alaska. *Geology* 15: 434-38.

Newhall, C. G., and Dzurisin, D. 1988. *Historical Unrest at Large Calderas of the World.* U.S. Geological Survey Bulletin 1855, v. 2.

Williams, Howel. 1958. *Landscapes of Alaska: Their Geologic Evolution.* Berkeley and Los Angeles: University of California Press.

Yount, E.; Miller, T. P.; Emanuel, R. P.; and Wilson, F. H. 1985. Eruption in an Ice-Filled Caldera, Mount Veniamin of, Alaska Peninsula. In *The Unites States Geological Survey in Alaska: Accomplishments during 1983*: ed. Bartsch-Winkler, S., and Reed, K. M., 59-60. U.S. Geological Survey Circular 945.

Chapter 23. Valley of the Ten Thousand Smokes

Alaska Travel Publications, Inc. 1974. *Exploring Katmai National Monument and the Valley of Ten Thousand Smokes.* Anchorage: Alaska Travel Publications, Inc.

Griggs, R. F. 1922. *The Valley of Ten Thousand Smokes.* Washington, D.C.: National Geographic Society.

Hildreth, Wes. 1983. The Compositionally Zoned Eruption of 1912 in the Valley of Ten Thousand Smokes, Katmai National Park, Alaska. *Journal of Volcanology and Geothermal Research* 18: 1-56.

Hildreth, Wes. 1987. New Perspectives on the Eruption of 1912 in the Valley of Ten Thousand Smokes, Katmai National Park, Alaska. *Bulletin of Volcanology* 49: 680-93.

Kienle, J., and Swanson, S. E. 1983. Volcanism in the Eastern Aleutian Arc: Late Quaternary and Holocene Centers, Tectonic Setting and Petrology. *Journal of Volcanology and Geothermal Research* 17: 393-432.

Martin, G. C. 1913. The Recent Eruption of Katmai Volcano in Alaska. *National Geographic Magazine* 24 :131-81.

Snyder, G. L. 1954. *Eruption of Trident Volcano, Katmai National Monument, Alaska, February-June, 1953.* U.S. Geological Survey Circular 318.

Chapter 24. Volcanic Hazards in the West: An Overview

Alt, David, and Hyndman, D. W. 1989. *Roadside Geology of Idaho*. Missoula, Montana: Mountain Press Publishing Company.

Bacon, C. R. 1985. Implications of Silicic Vent Patterns for the Presence of Large Crustal Magma Chambers. *Journal of Geophysical Research* 90 (B13): 11,243-52.

Blong, R. J. 1984. *Volcanic Hazards: A Sourcebook on the Effects of Eruptions*. Sydney, Australia: Academic Press.

Boyd, F. R., ed. 1984. *Explosive Volcanism, Inception, Evolution, and Hazards*. National Research Council Studies in Geophysics, Washington, D.C.: National Academy Press.

Bullard, Fred. 1984. *Volcanoes of the Earth*. 2nd. rev. ed. Austin, Texas: University of Texas Press.

Cameron, K. A., and Pringle, P. T. 1987. A Detailed Chronology of the Most Recent Eruptive Period at Mount Hood, Oregon. *Geological Society of America Bulletin* 99: 845-51.

Crandell, D. R. 1971. *Postglacial Lahars from Mount Rainier Volcano, Washington*. U.S. Geological Survey Professional Paper 677.

Crandell, D. R. 1973. *Potential Hazards from Future Eruptions of Mount Rainier, Washington*. U.S. Geological Survey Miscellaneous Geologic Investigations, Map I-836.

Crandell, D. R.; Mullineaux, D. R.; and Miller, C. Dan. 1979. Volcanic Hazards Studies in the Cascade Range of the Western United States. In *Volcanic Activity and Human Ecology*. ed. Sheets, P. D., and Grayson, D. K., 195-217. New York: Academic Press.

Crandell, D. R., and Nichols, D. R. 1987. *Volcanic Hazards at Mount Shasta, California*. Washington, D.C.: U.S. Geological Survey.

Decker, Robert; Wright, T. L.; and Stauffer, P. H., eds. 1987. *Volcanism in Hawaii*, U.S. Geological Survey Professional Paper 1350, 2 vols. Washington, D.C.: Government Printing Office.

Harris, Stephen L. 1988. *Fire Mountains of the West: The Cascade and Mono Lake Volcanoes*. Missoula, Montana: Mountain Press Publishing Co.

Hoblitt, R. P.; Miller, C. D.; and Scott, W. E. 1987. *Volcanic Hazards with Regard to Siting Nuclear Power Plants in the Pacific Northwest*. U.S. Geological Survey Open-File Report 87-297.

Martin, Roger C., and Davis, J. F. 1982. *Status of Volcanic Prediction and Emergency Response Capabilities in Volcanic Hazard Zones of California*. California Division of Mines and Geology, Special Publication 63.

Miller, C. Dan. 1980. *Potential Hazards from Future Eruptions in the Vicinity of Mount Shasta Volcano, Northern California*. U.S. Geological Survey Bulletin 1503.

Miller, C. Dan. 1985. Holocene Eruptions of the Inyo Volcanic Chain, California—Implications for Possible Eruptions Long Valley Caldera. *Geology* 13: 14-17.

Miller, C. D. 1989. *Potential Hazards from Future Volcanic Eruptions in California*. U.S. Geological Survey Bulletin 1847.

Miller, C. Dan; Mullineaux, D. R.; Crandell, D. R.; and Bailey, R. A. 1982. *Potential Hazards from Future Volcanic Eruptions in the Long Valley-Mono Lake Area, East-Central California and Southwest Nevada—A Preliminary Assessment*. U.S. Geological Survey Circular 877.

Miller, T. P., and Smith, R. W. 1987 Late Quaternary Caldera-Forming Eruptions in the Eastern Aleutian Arc, Alaska. *Geology* 15: 434-38.

Newhall, C. G., and Dzurisin, D. 1988. *Historical Unrest at Large Calderas of the World*. U.S. Geological Survey Bulletin 1855, vol. 2.

Rundle, J. B., and Hill, D. P. 1988. The Geophysics of a Restless Caldera—Long Valley, California. *Annual Review of Earth Planet. Science* 16: 251-71.

Sheets, P. D., and Grayson, D. K., eds. 1979. *Volcanic Activity and Human Ecology*. New York: Academic Press.

Sieh, K., and Bursick, M. 1986. Most Recent Eruptions of the Mono Craters, Eastern Central California. *Journal of Geophysical Research* 92 (B3) :2721-46.

Tilling, R.I. 1989. Volcanic Hazards and Their Mitigation: Progress and Problems. *Reviews of Geophysics*. 27 (2): 237-69.

Tilling, R. I., ed. 1989. *Volcanic Hazards: Short Course in Geology vol. 1*. Washington, D.C.: American Geophysical Union.

**Chapter 25.   A Look Back Into the Ice Age:
Glacier Bay National Park, Alaska**

Barnes, D. F., and Watts, R. D. 1977. *Geophysical Surveys in Glacier Bay National Monument*. U.S. Geological Survey Circular 751-B:93-95.

Boehm, W. D. 1975. *Glacier Bay: Old Ice, New Land, Alaska Geographic* 3: 1-134.

Harris, David, and Kiver, E. P. 1985. *The Geologic Story of the National Parks and Monuments*. 4th ed. New York: John Wiley and Sons.

Miller, M. 1967. Alaska's Mighty Rivers of Ice. *National Geographic* 131 (2): 194-217.

National Park Service. 1983. *Glacier Bay: A Guide to Glacier Bay National Park and Preserve, Alaska*. National Park Handbook 123, Washington, D.C.: National Park Service.

Trabant, D. C. 1976. *Alaska Glaciology Studies*. U.S. Geological Survey Circular 733: 45-47.

Chapter 26.   Glacial Lake Missoula: The World's Largest Flood

Allen, J. E. 1984. *The Magnificent Gateway: A Layman's Guide to the Geology of the Columbia River Gorge*. 2nd ed. Portland: Timber Press.

Allen, J. E.; Burns, Marjorie; and Sargent, S. C. 1986. *Cataclysms on the Columbia: A Layman's Guide to the Features Produced by the Catastrophic Bretz Floods in the Pacific Northwest*. Portland: Timber Press.

Bretz, J H. 1923. The Channeled Scablands of the Columbia Plateau, *Journal of Geology* 31: 617-49.

Bretz, J H. 1969. The Lake Missoula Floods and the Channeled Scabland. *Journal of Geology* 77: 505-543.

Bretz, J H., and Neff, G. E. 1956. Channeled Scabland of Washington: New Data and Interpretations. *Geological Society of America Bulletin*. 957-1049.

Waitt, R. B., Jr. 1980. About Forty Last-Glacial Lake Missoula Jokulhlaups through Southern Washington. *Journal of Geology* 88: 653-79.

Waitt, R. B., Jr. 1983. *Tens of Successive, Colossal Missoula Floods at North and East Margins of Channeled Scabland.* U.S. Geological Survey Open-File Report 83-671.

Waitt, R. B., Jr. 1984. Periodic Jokulhlaups from Pleistocene Lake Missoula—New Evidence from Varved Sediment in Northern Idaho and Washington. *Quaternary Research* 22:46-58.

Weis, P., and Newman, W. L. 1971. *The Channeled Scabland of Eastern Washington: The Geologic Story of the Spokane Flood.* U.S. Geological Survey Pamphlet.

Chapter 27. Fire Under Ice: The Steam Caves
At Mount Rainier National Park

Crandell, D. R. 1983. *The Geologic Story of Mount Rainier.* U.S. Geological Survey Bulletin 1292.

Haines, Aubrey L. 1962. *Mountain Fever! Historic Conquests of Rainier.* Portland, Oregon: Oregon Historical Society.

Harris, D. V., and Kiver, E. P. 1985. *The Geologic Story of the National Parks and Monuments.* 4th ed. New York: John Wiley and Sons.

Kiver, E. P. 1978a. Geothermal Ice Caves and Fumaroles, Mount Baker Volcano, 1974-1977 [abst.]. *Geological Society of America, Abstracts with Programs* 10 (3): 112.

Kiver, E. P. 1978b. Mount Baker's Changing Fumaroles. *The Ore Bin* 40 (8): 133-145.

Kiver, E. P., and Mumma, M. P. 1971. Summit Firn Caves, Mount Rainier, Washington. *Science* 173: 320-22.

Kiver, E. P., and Steel, W. K. 1975. Firn Caves in the Volcanic Craters of Mount Rainier, Washington. *National Speleological Society Bulletin* 37 (1).

Molenaar, Dee. 1971. *The Challenge of Rainier.* Seattle: The Mountaineers.

Mullineaux, D. R.; Sigafoos, R. S.; and Hendricks, E. L. 1969. *A Historic Eruption of Mount Rainier, Washington.* U.S. Geological Survey Professional Paper 650-B: 315-318.

Chapter 28. Indian Myth and Geologic Reality:
The Bridge of the Gods

Allen, John E. 1984. *The Magnificent Gateway: A Layman's Guide to the Geology of the Columbia River Gorge.* 2nd ed. Portland: Timber Press.

Allen John E.; Burns, Marjorie; and Sargent, S. C. 1986. *Cataclysms on the Columbia: A Layman's Guide to the Features Produced by the Catastrophic Bretz Floods in the Pacific Northwest*. Portland: Timber Press.

Atwell, Jim. 1973. *"Tahmahnaw," The Bridge of the Gods*. Chicago: Adams Press.

Balch, Frederic Homer. 1890. *The Bridge of the Gods: A Romance of Indian Oregon*. Portland: Binfords and Mort.

Bunnell, Clarence O. 1935. *Legends of the Klickitats: A Klickitat Version of the Story of the Bridge of the Gods*. Portland: Metropolitan Press.

Clark, Ella. 1953. *Indian Legends of the Pacific Northwest*. Berkeley, California: University of California Press.

Lawrence, D. B., and Lawrence, E. G. 1958. Bridge of the Gods Legend, Its Origin, History and Dating. *Mazama* 11 (13): 33-41.

Minor, Rick. 1984. *Dating the Bonneville Landslide in the Columbia River Gorge*. Heritage Research Report No. 31. Eugene, Oregon: Heritage Research Associates.

Palmer, L. 1977. Large Landslides of the Columbia River Gorge, Oregon and Washington. *Geological Society of America, Reviews in Engineering Geology* 3: 69-83.

Strong, Emory. 1960. *Stone Age on the Columbia River*. Portland: Binfords and Mort.

Williams, Ira A. 1923. *The Columbia River Gorge, Its Geologic History Interpreted from the Columbia River Highway*. Mineral Resources of Oregon, Oregon Bureau of Mines and Geology.

Chapter 29. Geomythology: The Battles of Llao and Skell

Bacon, C. R. 1983. Eruptive History of Mount Mazama and Crater Lake Caldera, Cascade Range, U.S.A. *Journal of Volcanology and Geothermal Research* 18: 57-115.

Clark, Ella. 1953. *Indian Legends of the Pacific Northwest*. Berkeley, California: University of California Press.

Miller, C. Dan. 1980. *Potential Hazards from Future Eruptions in the Vicinity of Mount Shasta Volcano, Northern California*. U.S. Geological Survey Bulletin 1503.

Vitaliano, Dorothy B. 1976. *Legends of the Earth: Their Geologic Origins*. Secaucus, N.J.: The Citadel Press.

Williams, Howel. 1942. *The Geology of Crater Lake National Park, Oregon*. Carnegie Institution Publication 540.

Chapter 30. Cosmic and Geologic Violence: Learning to Cope with Chaos

Alvarez, L. W. 1983. Experimental Evidence that an Asteroid Impact Led to the Extinction of Many Species 65 Million Years Ago. *Proceedings of the National Academy of Science, U.S.A.* 80: 627-42.

Alvarez, W., and Muller, R. A. 1984. Evidence from Cater Ages for Periodic Impacts on the Earth. *Nature* 308: 718-20.

California Division of Mines and Geology. 1989. How Do You Prepare for Something Like a Magnitude 7 or 8 Earthquake? *California Geology* 47 (4): 85-86.

Gore, Rick. 1989. Extinctions. *National Geographic.* 662-99.

Grieve, R. A. F. 1982. *The Record of Impact on Earth: Implications for a Major Cretaceous / Tertiary Impact Event.* Geological Society of America Special Paper 190: 25-37.

Grieve, R. A. F., and Dence, M. R. 1979. The Terrestrial Cratering Record. *Icarus.* 38: 230-42.

Grieve, R. A. F.; Sharpton, V. L.; Goodacre, A. K.; and Garvin, J. B. 1985. A Perspective on the Evidence for Periodic Cometary Impacts on Earth. *Earth Planet. Science Letter* 76: 1-9.

Harris, Stephen L. 1985. Earthquake Hazards in the West. *American West* 22 (3): 28-36.

Harris, Stephen L. 1983. Volcanic Hazards in the West. *American West* 20 (6): 30-39.

Mader, George G., and Blair, M. L. 1987. *Living with a Volcanic Threat: Response to Volcanic Hazards Long Valley, California.* Portola Valley, California: William Spangle and Associates, Inc.

Mark, Kathleen. 1987. *Meteorite Craters.* Tucson: University of Arizona Press.

Petak, W. J., and Atkisson, A. A. 1982. *Natural Hazard Risk Assessment and Public Policy, Anticipating the Unexpected.* New York, Heidelberg, Berlin: Springer-Verlag.

Rampino, M. R. 1987. Impact Cratering and Flood Basalt Volcanism. *Nature* 32: 468.

Rampino, M. R., and Stothers, R. B. 1988. Flood Basalt Volcanism during the Past 250 Million Years. *Science* 241: 663-68.

Rampino, M. R., and Stothers, R. B. 1984a. Terrestrial Mass Extinctions, Cometary Impacts, and the Sun's Motion Perpendicular to the Galactic Plane. *Nature.* 308: 709-12.

Rampino, M. R., and Stothers, R. B. 1984b. Geological Rhythm and Cometary Impacts. *Science* 226: 1427-31.

Rampino, M. R.; Self, S.; and Fairbridge, R. W. 1979. Can Rapid Climatic Change Cause Volcanic Eruptions? *Science* 206: 826-29.

Schnell, Mary L., and Herd, D. G. 1983. *National Earthquake Hazards Reduction Program: Report to the United States Congress.* U.S. Geological Survey Series of Annual Reports.

Sheets, P. D., and Grayson, D. K., eds. 1979. *Volcanic Activity and Human Ecology.* New York: Academic Press.

Silver, L. T., and Schultz, P. H., eds. 1982. *Geological Implications of Impacts of Large Asteroids and Comets on the Earth.* Boulder, Colorado: Geological Society of America.

Smoluchowski, R.; Bahcall, J. N.; and Matthews, M. S., eds. 1986. *The Galaxy and the Solar System*. Tucson: University of Arizona Press.

Steinbrugge, Karl V. 1982. *Earthquakes, Volcanoes, and Tsunamis: An Anatomy of Hazards*. New York: Skandia America Group.

Tilling, Robert I. 1989. *Volcanic Hazards: Short Course in Geology, Volume I*. Washington, D.C.: American Geophysical Union.

Tilling, Robert I. 1989. Volcanic Hazards and Their Mitigation: Progress and Problems. *Review of Geophysics* 27 (2): 237-69.

Trefil, James. 1989. Craters, the Celestial Calling Cards. *Smithsonian* 20 (6): 80-93.

Warrick, Richard A. 1975. *Volcano Hazard in the United States: A Research Assessment*. Monograph No. KSF-RA-E-75-012. Boulder, Colorado: Institute of Behavioral Science.

Glossary

Active fault: a fault along which movement has occurred in historic or recent geologic time, typically within the last 10,000 to 2 million years, and along which future movement is expected.

Aftershock: an earthquake that follows a larger earthquake or main shock and originates at or near the focus of the larger quake.

Amplification: the increase in earthquake ground motion that may occur in seismic waves as they enter and travel underground through certain kinds of rock.

Amplitude: the maximum height of a wave crest or depth of a trough.

Andesite: a volcanic rock, usually light gray or brown in color, of intermediate composition, with a silica content ranging from about 54 to 62 percent.

Ash: fine particles of pulverized rock, up to about 0.1 inch in diameter, blown from a volcano.

Ash fall: a rain of ash from an eruption cloud.

Ash flow: an avalanche of hot volcanic ash and gases that can travel great distances at high speeds from an erupting vent. See Pyroclastic flow.

Basalt: a dark-colored, fine-grained volcanic rock rich in iron and magnesium and relatively poor in silica (less than 54 percent by weight). The most common volcanic rock, basalt forms the ocean floors, mid-ocean volcanic islands, great lava plateaus, including the Deccan and Columbia River, and parts of subduction zone volcanic belts.

Caldera: a large basin-shaped volcanic depression, by definition at least a mile in diameter. Calderas typically are formed by the collapse of a volcanic cone or other structure.

Cinders: a general term applied to vesicular, usually dark-colored rock fragments.

Cinder cone: a volcanic cone built entirely of loose fragmental rock by moderately explosive eruptions.

Continental drift: the theory that horizontal movement of the earth's outer plates causes the continents slowly to shift their positions relative to each other.

Convection currents: movements of material caused by difference in density, typically the result of differences in temperature.

Crater: the bowl or funnel-shaped hollow at or near the top of a volcano, through which volcanic ash, lava flows, and gas are ejected.

Dacite: a typically light-colored volcanic rock with a high silica content, 63 to 70 percent. Gas-rich dacite magmas are commonly highly explosive, while gas-poor dacites typically form thick, viscous tongues of lava.

Dike: a sheetlike body of igneous rock that cuts through, in a generally vertical direction, older rock formations.

Dome: a generally rounded protrusion of lava that, when erupted, was too viscous to flow far and instead piled up over the erupting vent to form a mushroom-shaped plug.

Dormant: literally, "sleeping," a volcano that is inactive at present but which is expected to erupt again.

Earthquake: the vibration of the earth caused by seismic waves traveling outward from a sudden break or rupture in rocks beneath the surface.

Elastic rebound: the sudden release of progressively stored strain in rocks, resulting in movement along a fault.

Elastic strain: strain in which a deformed body recovers its original shape after the stress is released.

Epicenter: the point on the earth's surface directly above the focus, or hypocenter, of an earthquake.

Epoch: a division of geologic time less than a period, such as the Pleistocene epoch of the Quaternary period.

Erosion: the physical removal of rock by wind, running water, landslides, or the flow of glacial ice.

Eruption: the geologic process by which ash, lava flows, and gas are ejected onto the earth's surface by volcanic activity. Eruptions vary in behavior from the quiet overflow of molten rock (effusive type) to the tremendously violent expulsion of fragmental material.

Extinct: the term describing a volcano that is not expected to erupt again.

Fault: a fracture or zone of fractures in bedrock along which differential movement has occurred.

Fault creep: slow movement occurring along a fault without causing an earthquake.

Fissure eruption: a volcanic eruption that occurs along a narrow fissure or line of closely-spaced fractures in the earth's crust.

Focus: the hypocenter, point within the earth at which an earthquake originates.

Fumarole: a vent or opening in the ground through which issue steam or other volcanic gases.

Geomythology: ancient oral traditions or myths that preserve memories of geologic events, such as earthquakes or volcanic eruptions.

Geothermal gradient: rate of temperature increase associated with increasing depth beneath the earth's surface, normally about twenty-five degrees centigrade per kilometer.

Geothermal power: power derived from harnessing and exploiting the heat energy of the earth, as by tapping the heat from a hot spring, geyser, or volcano.

Glacier: a large mass of ice, formed by the compaction and recrystallization of snow, which moves downslope because of its weight and gravity.

Granite: a coarse-grained igneous rock composed mostly of quartz and feldspar.

Ground water: the water that fills the cracks and pore spaces in the ground.

Holocene epoch: the last 10,000 to 12,000 years since the end of the Pleistocene epoch (Ice Age), the division of geologic time in which we now live.

Hot spot: a persistent heat source in the earth's upper mantle unrelated to plate boundaries, such as the mantle plumes underlying Yellowstone or Hawaii.

Ice sheet: a glacier covering a large area, several thousand square miles.

Igneous rock: from the Latin word for "fire," igneous refers to rocks derived from the solidification of molten rock or magma.

Isoseismal: contour lines drawn on a map to separate one level of seismic intensity from another.

Juan de Fuca plate: a relatively small segment of the Pacific Ocean plate that is being subducted beneath the margin of the North American plate.

Lava: magma erupted onto the earth's surface.

Lava fountain: a gas-charged spray of fluid magma shooting into the air above a fissure or other volcanic vent.

Liquefaction: the transformation of a granular soil to a liquefied state caused by strong earthquake shaking; it typically causes ground failure and/or subsidence.

Main shock: the largest earthquake in a series.

Magma: molten rock confined within the earth; when erupted on the surface it is called lava.

Magma chamber: underground reservoir of molten rock that feeds volcanic eruptions.

Magnitude: a measure of the energy released during an earthquake or volcanic eruption.

Mantle: that portion of the earth's interior lying between the core and the crust, a zone of hot plastic rock extending approximately 1800 miles beneath the surface. This is the region in which magma is generated.

Meteor: a fragment of material from the solar system that becomes visible only when it enters Earth's atmosphere and becomes incandescent, heated through friction.

Meteorite: meteor that strikes the earth's surface.

Microearthquake: an earthquake too small to be felt by humans but detectable on a seismometer.

Mid-oceanic ridge: an enormous submarine mountain range that extends around the globe.

Modified Mercalli scale: scale designed to express the intensities of earthquakes, determined by their effects, especially the degree of damage they inflict in given locations. The scale expresses increasing intensities by Roman numerals I to XII.

Moraine: a typically linear ridge of rock fragments deposited by a glacier.

Mudflow: a water-saturated mass of rock debris that travels downslope as a liquid under the pull of gravity, a common by-product of volcanic eruptions.

Normal fault: a vertical or steeply inclined fault along which the overhanging block above the fault has moved downward relative to the block below.

Oceanic crust: the basaltic crust, three to four miles thick, that forms the ocean floors.

P wave: the primary, or fastest, wave radiating away from an earthquake focus and consisting of a series of compressions and expansions of rock material.

Paroxysm: a violently explosive eruption of great magnitude.

Plate: a large rocky slab of the outer rind of the earth and upper mantle that slowly moves over the plastic mantle beneath it. Three to four miles thick under the oceans, plates are much thicker under the continents.

Plate tectonics: a theory holding that earth's rigid outer shell is broken into about a dozen large slabs or plates that are in constant motion. Concentrations of earthquake and volcanic activity occur at the boundaries between plates.

Pleistocene epoch: the time between about 1.8 million and 10,000 years ago during which large continental ice sheets repeatedly formed throughout the Northern hemisphere; the Ice Age.

Plume: term designating the column of magma rising within the earth's mantle to produce "hot spot" volcanoes.

Pumice: a highly porous volcanic rock fragment formed of glassy lava-foam blown into the air during an eruption.

Pyroclastic: from the Greek word, "fire broken," volcanic rock ejected in fragments.

Pyroclastic flow: an avalanche of incandescent rock fragments and hot gas that travels downslope as a heavy "fluid." See Ash flow.

Reverse fault: a sloping fault in which the crustal block above moves up the fault surface compared with the block below.

Right lateral fault: a fault in which the crustal block on the other side of the fault from an observer has moved to the right.

S wave: the secondary seismic wave, traveling more slowly than the P wave, and consisting of elastic vibrations that cause undulating or side-to-side ground movement.

Sea floor-spreading: term describing the movement of sea floor plates away from mid-oceanic ridges; basalt magma rises to fill new fissures and creates new oceanic crust. Plates moving sea floor toward the continents are subducted beneath continental margins.

Seismic wave: an elastic wave in the earth generated by an earthquake.

Seismograph: an instrument that records the degree of ground motion during an earthquake.

Shield volcano: a broad gently sloping volcanic structure built almost exclusively of thin lava flows. Named for their supposed resemblance to an Icelandic warrior's shield laid down flat with the convex side upward, shield volcanoes commonly form by the quiet effusion of fluid basaltic lava.

Silica: silicon dioxide, the chemical combination of silicon and oxygen, a primary constituent of volcanic rocks.

Silicic lava: term describing lava rich in silica (more than about 63 percent) and having a relatively low melting point (about 850 degrees Centigrade). Silicic magma typically forms a stiff, viscous mass that builds lava domes like Lassen Peak.

Slip: the motion of one side of a fault relative to the other.

Stratovolcano: also known as a composite cone, a volcano composed of both lava flows and fragmental material.

Stress: a measure of the forces acting on a body in units of force per unit area.

Strike: the direction or trend taken by a surface such as a fault plane as it intersects the horizontal.

Strike-slip fault: a fault on which movement is principally horizontal, parallel to the fault's strike.

Strong ground motion: the shaking of the ground near an earthquake source made up of large amplitude seismic waves of various types.

Subduction: the process by which sea floor sinks beneath the margin of a continent or island arc.

Subduction zone: the region of convergence of two tectonic plates, one of which sinks beneath the other, as where the Pacific plate is sliding beneath the Pacific Northwest coast of North America.

Surface wave: a seismic wave that follows the earth's surface only; it consists of Rayleigh and Love waves, which cause both rolling and zig-zag ground movement highly damaging to many structures.

Tectonics: the study of major structural features of the earth and the processes that create them.

Tephra: rock fragments of all sizes thrown into the air above a volcano.

Thrust fault: a reverse fault in which the dip of the fault plane is at a low angle to the horizontal.

Tsunami: a long ocean wave typically caused by sea floor displacements in an earthquake.

Unconsolidated: term referring to rock particles that are loose, separate, or unattached to each other. Unconsolidated soils typically slide, slump, and shake more severely than bedrock during an earthquake.

Vent: a volcanic opening in the earth's surface through which volcanic ash, lava flows, and/or gas are emitted.

Volcano: the hill or mountain built by ash and/or lava flows ejected from a volcanic vent.

Wavelength: the distance betweeen two successive crests or troughs of a wave.

Index

Cascadia subduction zone, 19, 72, 74-76, 78, 96-97, 111, 207, 212
Castle Rock, 107
catastrophists, 195
Chain of Craters Road, 147, 151,
Channeled Scabland, 194, 196,
chaos, 1, 4, 201, 221-224, 227
Chaos Crags, 119
Chaos Jumbles, 119
Charleston, 91-92, 103
Chicago, 90, 91,
Cholame Valley, 38
Chugach Mountains, 66
Cincinnati, 90
Cinder Cone, 119
cinder cones, 139, 177-80
city hall, San Francisco, 41
Clark Fork River, 194, 198
climate, effect of meteorite impacts on, 158-59, 222
    effect of volcanic eruptions on, 128, 181
    see also Ice Age
Cold Bay, 164
Columbia River, 107, 114, 127, 196, 199, 201, 208, 209, 211-13
Columbia River Gorge, 77, 194, 198, 199, 212
Columbia River Plateau, 2, 127-34, 135, 137, 139, 194, 198
Colvig, William M., 215, 219
Commander Gap, 71
comets, 222, 223, see also meteorite impacts
composite cones, 111
Cook Inlet, 163, 164, 166
Cordilleran Ice Sheet, 194, 198
Cordova, 66
core, of earth, 17-18
cosmos, 224
Cowlitz River, 107, 112
Crater Lake, 109, 111, 116, 214, 215, 218, 220, see also Mazama, Mount
Crater Rock, 114, 115, 213
Craters of the Moon National Monument, 139
Creede caldera, 186

Crescent City, 67
Cretaceous, 132,
Cretaceous-Tertiary boundary, 223
curtain of fire, 151

Daly City, 100, 101, 224
Deccan Plateau, 129, 132, 158
Death Valley, 79-81, 84-85
Deschutes River, 115, 196
Detroit, 88
Devil's Golf Course, 80
Diller, J. S., 219
Drift River oil terminal, 163
Drift River valley, 163
Dry Falls, 197
Dunsmuir, 116

Eagle Creek formation, 211
earth, interior of, 17-18
Eliot Glacier, 114
Embarcadero Freeway, 25
Emmons Glacier, 202, 205, 206, 207
Emperor Seamounts, 154, see also Hawaii-Emperor chain
Enumclaw, 111
epicenter, 5
Eureka, 75
exotic terranes, 21-23
Exxon oil spill, 68

Fairweather, Mount, 189
Fairweather Range, 189
faults, 8
    Calaveras, 225
    Hayward, 47, 48, 49, 53-56, 63, 98
    Newport-Inglewood, 47, 48, 59-63, 98
    Owens Valley, 44-45
    San Andreas, 10-11, 36, 37-39, 47-48, 49, 51-52, 57-59, 63, 98-99, 100, 101
fires, following earthquakes, 48, 56, 61, 62
    San Francisco, 25, 27, 41-43, 51, 57

256